Conference of Theoretical Physics and Nonlinear Phenomena 2016

Journal of Physics: Conference Series Volume 856

Jakarta, Indonesia
4 October 2016

ISBN: 978-1-5108-4192-5
ISSN: 1742-6588

Printed from e-media with permission by:

Curran Associates, Inc.
57 Morehouse Lane
Red Hook, NY 12571

Printed by Curran Associates, Inc. (2017)

For permission requests, please contact the Institute of Physics
at the address below.

Institute of Physics
Dirac House, Temple Back
Bristol BS1 6BE UK

Phone: 44 1 17 929 7481
Fax: 44 1 17 920 0979

techtracking@iop.org

Additional copies of this publication are available from:

Curran Associates, Inc.
57 Morehouse Lane
Red Hook, NY 12571 USA
Phone: 845-758-0400
Fax: 845-758-2633
Email: curran@proceedings.com
Web: www.proceedings.com

Table of contents

Volume 856

Conference on Theoretical Physics and Nonlinear Phenomena 2016
4 October 2016, Jakarta, Indonesia

Accepted papers received: 25 May 2017
Published online: 2 June 2017

Preface

Papers

IOP Conf. Series: Journal of Physics: Conf. Series **856** (2017) 011001 doi:10.1088/1742-6596/856/1/011001

Preface

The Conference on Theoretical Physics and Nonlinear Phenomena is an annual meeting on theoretical physics and nonlinear phenomena in Indonesia. The current event was the 13th meeting, held in Jakarta, Indonesia on 4 October 2016. The first conference of this series, then named Workshop on Theoretical Physics, was initiated in 2004.

The Conference was attended by 26 participants. There were 4 invited talks and 11 contributed talks presented in the conference, covering the general vast area of theoretical physics and nonlinear phenomena.

The original spirit of founding organizers of the 1st Workshop on Theoretical Physics (WTP) was to create an annual, regular forum for the community of physicists in Indonesia working in the fields of fundamentals areas of physics. We are happy to report that in the twelve years since the first workshop, the original spirit has remained strong as ever. Of particular note is the emergence of bright undergraduate and master students presenting their works; and the presentation by young scientists freshly returned from their doctoral studies and postdoctoral appointments abroad.

We would like to deeply thank all the parties involved in organizing the conference, the session chairs, and the Research Center for Physics, Indonesian Institute of Sciences who acted as the host institution for this conference.

The complete contributions of the conference will be available as IOP Conference Proceedings.

Co-chairs of the Conference

Ardian N. Atmaja
J. Julio

Peer review statement

All papers published in this volume of *Journal of Physics: Conference Series* have been peer reviewed through processes administered by the proceedings Editors. Reviews were conducted by expert referees to the professional and scientific standards expected of a proceedings journal published by IOP Publishing.

IOP Conf. Series: Journal of Physics: Conf. Series **856** (2017) 012001 doi:10.1088/1742-6596/856/1/012001

Role of the nucleon resonances with different spins in the photoproduction of kaon on the nucleon

T Mart and S Clymton

Departemen Fisika, FMIPA, Universitas Indonesia, Depok 16424, Indonesia

E-mail: terry.mart@sci.ui.ac.id

Abstract. We have studied the role of nucleon resonances with different spin numbers in the kaon photoproduction process $\gamma p \to K^+\Lambda$ by using an isobar, where the included nucleon resonances have spins up to 9/2. The model was fitted to around 7400 experimental data points. It is found that the most important resonances in the $\gamma p \to K^+\Lambda$ reaction are those with spin 3/2, followed by spin 5/2. Contribution of the higher spin resonances is found to be less important in this process.

1. Introduction

The electromagnetic production of meson has been known as a very useful tool to study the nucleon resonances. Most of these resonances decay into the pion-nucleon channels, where theoretical predictions can be made with a high accuracy. However, a number of these resonances have only sizable decay probability to the kaon-hyperon channels, according to the constituent quark models [1]. In fact, the number of nucleon resonances predicted by the constituent quark models [1, 2] is considerably more than what has been observed in the $\pi N \to \pi N$ scattering experiments. This phenomenon was later known as the "missing resonances" [3]. For instance, the $N(2030)S_{11}$, $N(2145)S_{11}$, and the $N(2195)S_{11}$, were predicted to mostly decay into the $K\Sigma$ channel rather than to the πN channels [1]. Therefore, up to now these states have not been observed by the Particle Data Group [4], whose main tool for observing and extracting the properties of resonance states is the $\pi N \to \pi N$ channels.

To accurately analyze the resonance states in the kaon channels we need a reliable tool. To this end the most often used tool is the isobar model, which is based on the field theoretic or multipoles approach. Basically, the first one is constructed from the suitable Feynman diagrams, where the unknown parameters in the electromagnetic and hadronic vertices can be extracted by fitting the predicted observables to experimental data. Note that the purpose of the present paper is to describe the progress of our work in this approach. The second method, i.e., the multipoles one, can be performed by parameterizing the resonance properties with the help of the Breit-Wigner form (see [5–8] for examples).

2. The isobar model

The isobar model used in the present work is based on our previous report [9]. This model was constructed by using the field theoretic approach and consists of the standard s-, u- and t-channel for the Born terms along with the $K^{*+}(892)$, $K_1(1270)$ vector mesons and the $\Lambda(1600)P_{01}$ and $\Lambda(1810)P_{01}$ hyperon resonances. The two hyperon resonances have been shown to play an important role in increasing the values of the leading coupling constants as well as the hadronic form factor cutoff [10]. To account for the hadronic structures, hadronic form factors are included in all hadronic interactions. To restore the gauge invariance of the amplitude, which is destroyed by the inclusion of the form factors, we exploit the Haberzettl prescription [11]. The effect of different types of hadronic form factors has been also investigated and the report can be found in [12,13]. Another issue related to the contribution of the nucleon resonances below the threshold energy has been explained in [14].

The nucleon resonances used in the model are obtained from the list given in the Review of Particle Properties of the Particle Data Group (PDG) [4]. For the best model with the smallest χ^2 we include all resonances with spins up to 9/2, i.e., the $N(1440)P_{11}$, $N(1520)D_{13}$, $N(1535)S_{11}$, $N(1650)S_{11}$, $N(1675)D_{15}$, $N(1680)F_{15}$, $N(1700)D_{13}$, $N(1710)P_{11}$, $N(1720)P_{13}$, $N(1860)F_{15}$, $N(1875)D_{13}$, $N(1880)P_{11}$, $N(1895)S_{11}$, $N(1900)P_{13}$, $N(2000)F_{15}$, $N(2060)D_{15}$, $N(2120)D_{13}$, $N(1990)F_{17}$, $N(2190)G_{17}$, $N(2220)H_{19}$, and $N(2250)G_{19}$.

Since the present analysis relies on the single channel method, it is obvious that the unitarity might be violated. Therefore, a correction at the tree-level is necessary. For this purpose we use the energy-dependent widths with partial branching fractions in the resonance propagators [15].

The model has been fitted to all available experimental data, consisting of about 7400 data points, by minimizing the value of

$$\frac{\chi^2}{N_{\rm dof}} \equiv \frac{1}{(N_{\rm data} - N_{\rm par})} \sum_{i=1}^{N_{\rm data}} \left(\frac{\sigma_i^{\rm exp} - \sigma_i^{\rm th}}{\Delta\sigma_i^{\rm exp}} \right)^2 , \tag{1}$$

where $N_{\rm data}$ and $N_{\rm par}$ refer to the numbers of experimental data and fitting parameters, $\sigma_i^{\rm exp}$ and $\Delta\sigma_i^{\rm exp}$ denote the i-th experimental observable (total cross section, differential cross section, single or double polarization observable) and its error bar, and $\sigma_i^{\rm th}$ is the corresponding theoretical calculation. Note that the $N_{\rm dof}$ represents the number of degrees of freedom. A comparison of small part of these data with our best model can be found in our recent report [9].

3. Results and discussion

Since the number of coupling constants of the nucleon resonances used in this analysis is very large, we only present the background parameters of the models along with the corresponding $\chi^2/N_{\rm dof}$ values in table 1, where different models refer to different excluded nucleon resonances. In the second column of table 1 we list the background parameters of the best model, where all nucleon resonances with different spin values are included. In other columns of table 1 the spin number of excluded nucleon resonances is indicated in the header of the table. To have a better view of the role of these resonances the values of $\chi^2/N_{\rm dof}$ listed in table 1 is plotted in figure 1.

It is obvious that the exclusion of spin-3/2 nucleon resonances leads to the largest χ^2, which indicates that the spin-3/2 nucleon resonances play the most important role in the $\gamma p \to K^+\Lambda$ process. This finding is consistent with our previous result, which shows that the $N(1720)P_{13}$ and $N(1900)P_{13}$ states are the responsible resonances for explaining the first and second peaks of the $K^+\Lambda$ cross section [16].

The second important resonances are those with spin 5/2. The importance of these resonances has been also highlighted in our recent report [9]. It is quite surprising that the role of spin-1/2 resonances is less important than that of spin-3/2 and -5/2 resonances. Presumably, this originates from the simple form of the spin-1/2 resonance amplitude. Furthermore, the spin-1/2 resonance has only one parameter for the coupling constant. Obviously, the amplitude is less flexible than those of spin-3/2 or -5/2 resonances, where the number of parameters is two, at least, to fit the experimental data. We note, however, that the spin-1/2 state $N(1650)S_{11}$ is

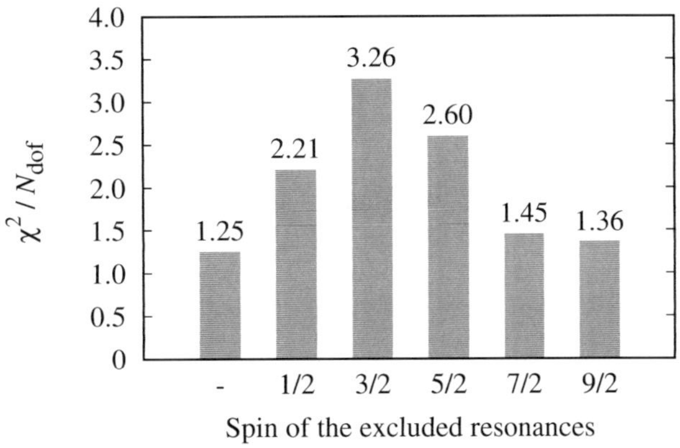

Figure 1. Values of $\chi^2/N_{\rm dof}$ if the nucleon resonances with certain spin number are excluded from the model. The best model with all nucleon resonances included is indicated by the left bar with $\chi^2/N_{\rm dof} = 1.25$.

very important in kaon photoproduction [5–7]. Therefore, a more detailed study of the spin-1/2 resonance role is urgently required in the future studies of kaon photoproduction in order to clarify this situation.

The role of higher spin resonances (7/2 and 9/2) is certainly small. This result is as expected, since it requires high energy and momentum to align the spins of quarks to produce baryons with high spin values. Consequently, the probability of finding such resonances is small in our process and their role is also less important.

For completeness, we present the behavior of different models discussed above in figure 2 in term of the calculated total cross sections as a function of the total center of momentum energy

Table 1. List of the background parameters, the hadronic form factor cutoffs, and the $\chi^2/N_{\rm dof}$ obtained by including all resonances (Best) and excluding nucleon resonances with a specific spin number. Notation of the parameters is as in our previous studies [3, 9].

Parameters	Best	Spin-1/2	Spin-3/2	Spin-5/2	Spin-7/2	Spin-9/2
$g_{K\Lambda N}/\sqrt{4\pi}$	−3.00	−3.00	−4.40	−4.40	−3.00	−3.00
$g_{K\Sigma N}/\sqrt{4\pi}$	0.90	0.90	1.30	1.30	1.30	0.90
$G^V_{K^*}/4\pi$	−0.18	−0.50	−0.00	−0.09	0.05	−0.23
$G^T_{K^*}/4\pi$	0.72	0.83	0.11	0.01	0.18	0.72
$G^V_{K_1}/4\pi$	−0.63	0.12	−0.02	0.01	0.49	−0.70
$G^T_{K_1}/4\pi$	−2.94	−2.84	−0.43	−0.49	0.26	−2.80
$G_{\Lambda(1600)}/4\pi$	−7.19	−10.0	−9.23	−6.63	7.62	−7.78
$G_{\Lambda(1810)}/4\pi$	10.0	3.49	−10.0	10.0	−10.0	10.0
$\Lambda_{\rm B}$(GeV)	0.70	0.70	1.11	0.79	0.86	0.70
$\Lambda_{\rm R}$(GeV)	1.18	1.17	0.98	1.12	1.25	1.17
$\theta_{\rm had}$(deg)	90.0	115	125	90.0	111	90.0
$\phi_{\rm had}$(deg)	0.01	90.0	159	0.00	141	0.08
$\chi^2/N_{\rm dof}$	1.25	2.21	3.26	2.60	1.45	1.36

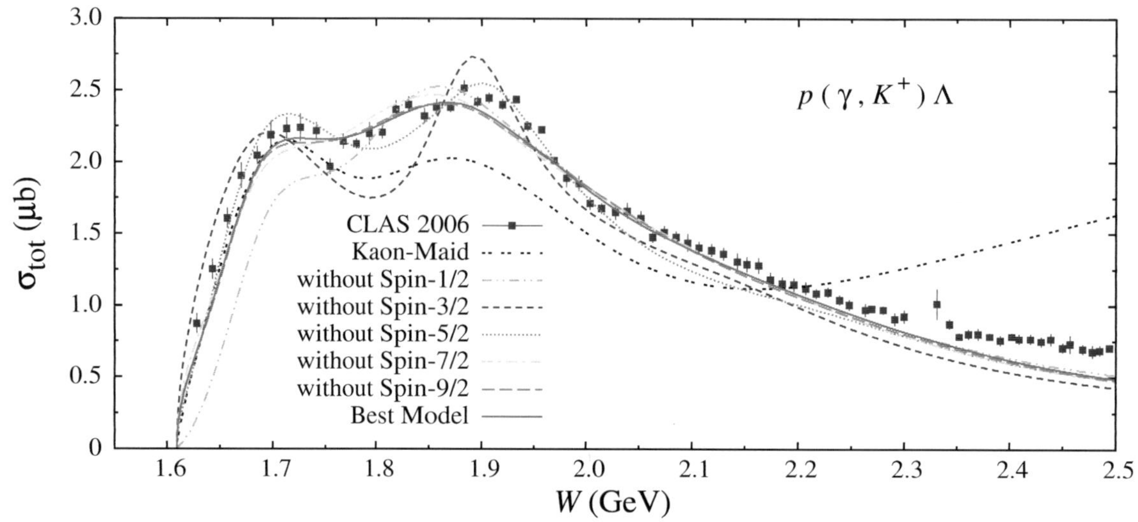

Figure 2. Total cross section calculated by using all resonances (Best Model) compared with those obtained by excluding nucleon resonance with a specific spin. Experimental data obtained from the CLAS Collaboration (solid squares [17]) and prediction of Kaon-Maid [3] are also shown for comparison. Note that the experimental data shown in this figure are only for comparison and were not included in the fitting process.

W, where the corresponding experimental data are also shown for comparison. Figure 2 clearly reveals that the role of the spin-1/2 resonances is very decisive in the low energy region. Without the spin-1/2 resonances it is impossible to explain the first peak of the cross section, as we have previously discussed. On the other hand, figure 2 also displays that without spin-3/2 resonances the calculated cross section deviates significantly from the experimental data.

4. Summary and conclusion

We have investigated the role of nucleon resonances with different spin values in the kaon photoproduction process $\gamma p \to K^+\Lambda$. We have found that the nucleon resonances with spin 3/2 play the most important role in this process, followed by the spin-5/2 and spin-1/2 resonances. The higher spin resonances are found to be less important for this process.

Acknowledgments

This work was partly supported by University of Indonesia.

References

[1] Capstick S and Roberts W 1998 *Phys. Rev.* D **58** 074011
[2] Capstick S and Roberts W 1994 *Phys. Rev.* D **49** 4570
[3] Mart T and Bennhold C 1999 *Phys. Rev.* C **61** 012201
[4] Olive K A *et al* 2014 *Chin. Phys.* C **38** 090001
[5] Mart T and Sulaksono A 2006 *Phys. Rev.* C **74** 055203
[6] Mart T 2010 *Phys. Rev.* C **82** 025209
[7] Mart T 2011 *Phys. Rev.* C **83** 048203
[8] Mart T 2014 *Phys. Rev.* C **90** 065202
[9] Mart T, Clymton S and Arifi A J 2015 *Phys. Rev.* D **92** 094019
[10] Mart T and Nurhadiansyah N 2013 *Few-Body Syst.* **54** 1729
[11] Haberzettl H, Bennhold C, Mart T and Feuster T 1998 *Phys. Rev.* C **58** 40
[12] Mart T and Sari A K 2013 *Mod. Phys. Lett.* A **28** 1350054

IOP Conf. Series: Journal of Physics: Conf. Series **856** (2017) 012001 doi:10.1088/1742-6596/856/1/012001

[13] Syukurilla L and Mart T 2015 *Int. J. Mod. Phys.* E **24** 1550008
[14] Mart T 2013 *Phys. Rev.* C **87** 042201
[15] Lee F X, Mart T, Bennhold C, Haberzettl H and Wright L E 2001 *Nucl. Phys.* A **695** 237
[16] Mart T and Kholili M J 2012 *Phys. Rev.* C **86** 022201
[17] Bradford R *et al* 2006 *Phys. Rev.* C **73** 035202

IOP Conf. Series: Journal of Physics: Conf. Series **856** (2017) 012002 doi:10.1088/1742-6596/856/1/012002

Experimental study of matter under the influence non-abelian gauge theory

Suharyo Sumowidagdo

Research Center for Physics, Indonesian Institute of Sciences, Serpong, Banten, Indonesia

E-mail: Suharyo.Sumowidagdo@lipi.go.id

Abstract. Physical phenomena, which are observed in our daily life, are dominated by electromagnetic interaction, an abelian gauge theory. In the region of highest experimentally reachable temperature and density, the strong interaction or QCD dominates. In this region, nuclear matter displays exotic properties which arise due to the non-abelian nature of QCD. In this talk I will illustrate some theoretical predictions of QCD using recent results from the ALICE Experiment.

1. Asymptotic freedom in quantum chromodynamics

Quantum chromodynamics (QCD) is the currently accepted theory that describes the phenomena of strongly interacting particles. It is a non-abelian gauge theory based on $SU(3)$ symmetry group. A particular feature of non-abelian gauge theory is asymptotic freedom [1, 2]. The origin of this feature is that the running behavior of coupling constant in non-abelian gauge theories is not only due to the fermion loops, but also due to gauge boson loops. The two kinds of diagrams produce opposite effects: the fermion loop diagrams tend to screen the pole charge, while the gauge boson loop diagrams tend to augment (antiscreen) the pole charge.

In QCD, the lowest-order beta function for the strong coupling α_3 is given by

$$\beta_3(\alpha) = \frac{\alpha_3^2}{\pi}\left(-\frac{33}{6}+\frac{n_f}{3}\right) \tag{1}$$

where n_f is the number of quark flavors. If the beta function is negative, the theory is asymptotically free. For QCD, it is asymptotically free as long as the number of quark flavors is less than 16. Experimentally, the number of quark flavors is 6, thus QCD is asymptotically free.

The thermodynamic property of strongly interacting matter is shown in the QCD phase diagram as shown in figure 1. A particular region in the phase diagram is the quark-gluon plasma region, which is the QCD analogy of electromagnetic plasma. In this region, the quarks are deconfined and together with the gluons, they form the quark-gluon plasma.

In order to study the property of matter under the influence of QCD in the asymptotic (short-distance) region, nuclei are accelerated and are collided at high energy to create highly compressed nuclear matter, as shown in figure 2. The quark-gluon plasma is behaving like a perfect liquid of strongly interacting matter. As this drop of perfect liquid expands and cools down, it hadronizes into the more well-known hadrons (mesons and baryons), which can be observed experimentally.

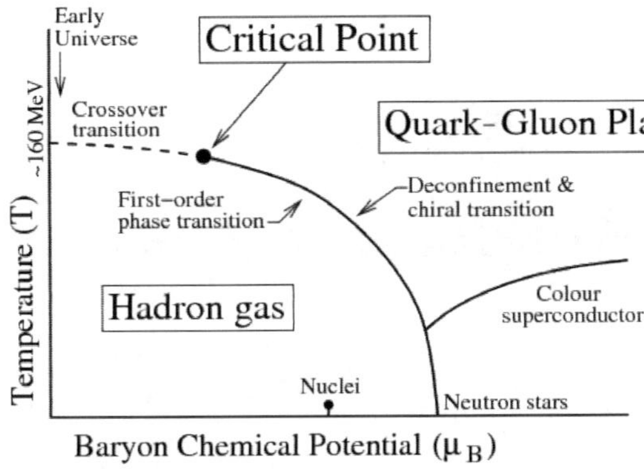

Figure 1. The thermodynamics phase diagram of QCD matter. Figure is taken from [3].

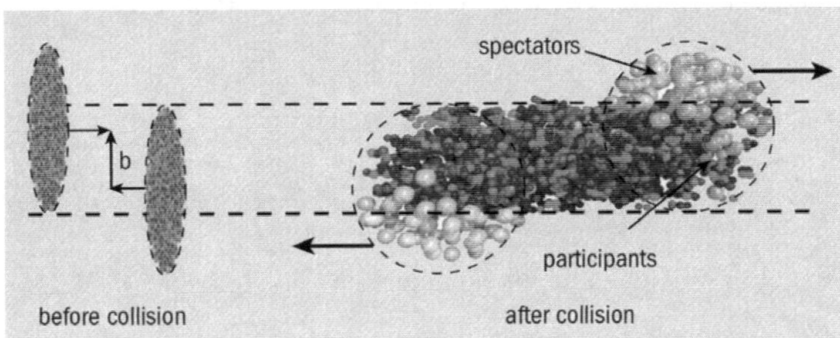

Figure 2. Collisions of two heavy nuclei in relativistic heavy ion collisions. The collisions are not always head-on, thus sometimes some of the nucleons become spectators while the rest become participants. Figure is taken from [4].

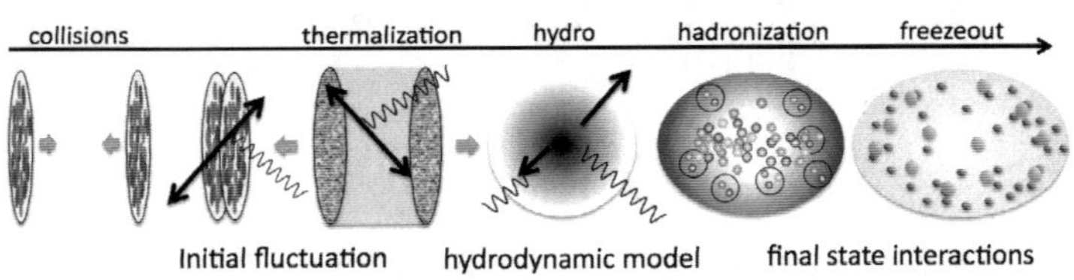

Figure 3. The expanding quark-gluon plasma cools down and hadronizes into the known hadrons. Figure is taken from [5].

2. Phenomenology and experimental results of heavy-ion collisions physics

One of the dedicated experiments to study quark-gluon plasma is A Large Ion Collider Experiment (ALICE) [6] at the Large Hadron Collider (LHC) [7]. The ALICE detector, shown in figure 4 is the only detector at the LHC that is designed specifically to optimally study the phenomena of quark-gluon plasma. The core of ALICE consists of tracking detectors, which provide three-dimensional tracking capability in busy environments, and particle identification detectors, which is used to separate protons, pions and kaons.

The ALICE detector is also capable of studying physical phenomena in proton-proton

IOP Conf. Series: Journal of Physics: Conf. Series **856** (2017) 012002 doi:10.1088/1742-6596/856/1/012002

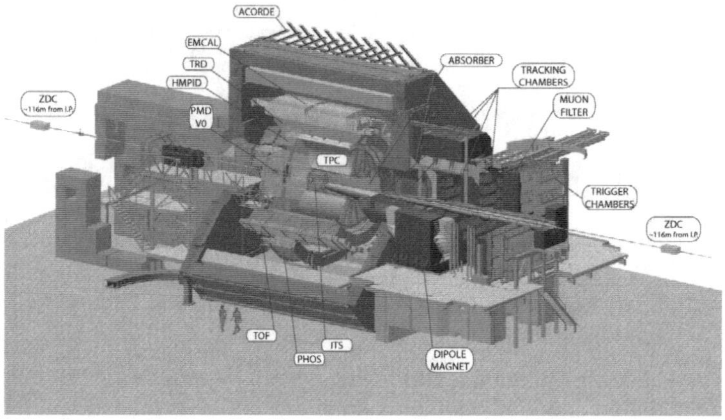

Figure 4. The ALICE Detector. Figure taken from [6].

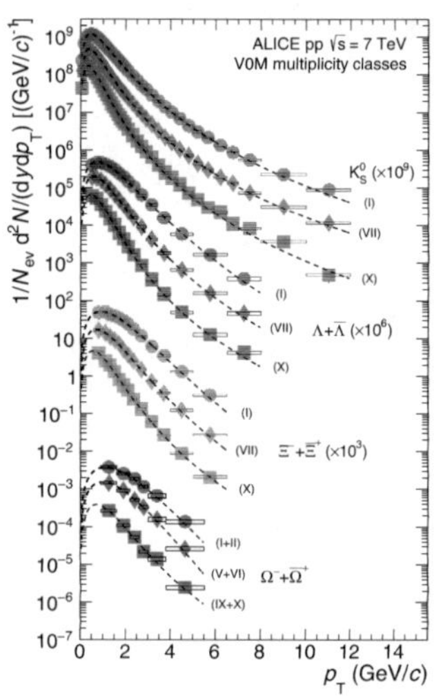

Figure 5. Production spectra of strange meson and bayons in proton-proton collisions. Figure from Ref [8].

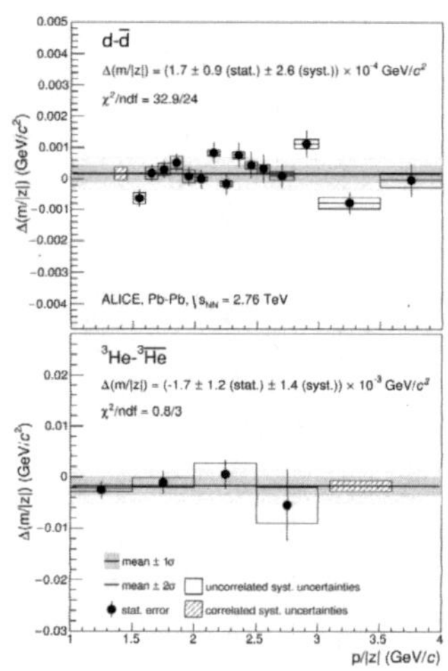

Figure 6. The mass difference of nuclei-antinuclei measured by ALICE. Figure from Ref [9].

collisions, in complementary of ATLAS and CMS. In 2016, ALICE reported the observation of enhancement in strangeness production measured in the spectra of kaons, Λs, Ξs and Ωs in proton-proton collisions as shown in figure 5 [8]. This enhancement is interpreted as formation of quark-gluon plasma in proton-proton collisions, and thus opens new possibilities to study quark-gluon plasma in a system smaller compared to ion-ion collisions.

The environment of heavy-ion collisions also provides the opportunity to observe and measure the property of antinuclei. ALICE has reported the first measurements of mass differences between deuteron-antideuteron, and between He^3 and $\bar{\mathrm{He}}^3$ as shown in figure 6 [9].

IOP Conf. Series: Journal of Physics: Conf. Series **856** (2017) 012002 doi:10.1088/1742-6596/856/1/012002

3. The future of ALICE Experiment

ALICE is planned to undergo major upgrade during the Long-Run Shutdown 2 (LS2) of the LHC [10]. The inner tracking systems will be completely rebuilt [11], the time projection chamber readout pads will be replaced with new pads made using gaseous electron multiplier [12], and the data acquisition system and offline computing will be merged into one [13].

ALICE is expected to run for a data-taking period of 6 years beyond LS2, providing even more opportunities for probing the property of strongly interacting matter under QCD. The future is surely rich with new possibilities of physics topics to be studied with ALICE.

References

[1] Gross D J and Wilczek F 1973 *Phys. Rev. Lett.* **30** 1343
[2] Politzer H D 1973 *Phys. Rev. Lett.* **30** 1346
[3] Bhalerao R S 2014 Relativistic heavy-ion collisions (*Preprint* `arXiv:1404.3294 [nucl-th]`)
[4] Toia A 2013 *CERN Courier* **April** 31
[5] Nonaka C and Asakawa M 2012 *Prog. Theor. Exp. Phys.* **2012** 01A208
[6] Aamodt K *et al* 2008 *J. Instrum.* **3** S08002
[7] Evans L and Bryant P 2008 *J. Instrum.* **3** S08001
[8] Adam J *et al* 2016 Multiplicity-dependent enhancement of strange and multi-strange hadron production in proton-proton collisions at $\sqrt{s} = 7$ TeV (*Preprint* `arXiv:1606.07424 [nucl-ex]`)
[9] Adam J *et al* 2015 *Nature Phys.* **11** 811
[10] Abelev B *et al* 2014 *J. Phys.* **G41** 087001
[11] Abelev B *et al* 2014 *J. Phys.* **G41** 087002
[12] Abelev B *et al* 2013 Technical design report for the upgrade of the alice time projection chamber Tech. rep. URL `https://cds.cern.ch/record/1622286`
[13] Buncic P, Krzewicki M and Vande Vyvre P 2015 Technical design report for the upgrade of the online-offline computing system Tech. rep. URL `https://cds.cern.ch/record/2011297`

IOP Conf. Series: Journal of Physics: Conf. Series **856** (2017) 012003 doi:10.1088/1742-6596/856/1/012003

Adomian decomposition method used to solve the one-dimensional acoustic equations

Meta Dispini[1] and Sudi Mungkasi[2]

[1] Postgraduate Program in Mathematics Education, Faculty of Teacher Training and Education, Sanata Dharma University, Mrican, Tromol Pos 29, Yogyakarta 55002, Indonesia
[2] Department of Mathematics, Faculty of Science and Technology, Sanata Dharma University, Mrican, Tromol Pos 29, Yogyakarta 55002, Indonesia

E-mail: mdispini@yahoo.com, sudi@usd.ac.id

Abstract. In this paper we propose the use of Adomian decomposition method to solve one-dimensional acoustic equations. This recursive method can be calculated easily and the result is an approximation of the exact solution. We use the Maple software to compute the series in the Adomian decomposition. We obtain that the Adomian decomposition method is able to solve the acoustic equations with the physically correct behavior.

1. Introduction

Physical phenomena are interesting study for the last decades. Many methods have been developed successfully to solve partial differential equations that help to analyze real-world problems. One interesting real-world problem is wave propagation. This paper considers the one-dimensional wave equations, in particular the one-dimensional acoustic equations. Acoustic equations can be derived from the nonlinear elastic equations, as described by LeVeque [1].

In this paper, the acoustic equations are solved using the Adomian decomposition method. Series in the Adomian decomposition method converges to the exact solution. The method attacks the problem in a straightforward way without any transformation formulas [2]. The method implements an integral operator to solve differential equations. When the exact solution cannot be expressed explicitly, the Adomian decomposition method results in approximations of the exact solution [3–6]. More terms involved in the Adomian decomposition solution series result in a more accurate approximation. To our knowledge, our work is the first in applying the Adomian decomposition method to solve the acoustic equations.

The rest of the paper is organized as follows. First, we write the problem description. Then we present the Adomian decomposition method. After that, we provide some computational results and discussion. Finally, concluding remarks are drawn.

2. Problem description

In this section, we describe the problem (the mathematical model) that we want to solve. Starting from the general model, we simplify the model into the simplest form of acoustic equations.

The general form of the acoustic equations are [7, 8]

$$p_t + K(x)\, u_x = 0, \tag{1}$$

$$\rho(x)\,u_t + p_x = 0. \tag{2}$$

Here $p(x,t)$ denotes the pressure, $u(x,t)$ represents the velocity, x is the one-dimensional space variable, and t is the time variable. In addition, $K(x)$ is the bulk modulus of compressibility and $\rho(x)$ is the density. We use derivative operators $p_t = \partial p/\partial t$, $p_x = \partial p/\partial x$, $u_t = \partial u/\partial t$ and $u_x = \partial u/\partial x$.

Taking $K(x) = 1$ and $\rho(x) = 1$, we obtain the acoustic equations in the simplest form

$$p_t + u_x = 0, \tag{3}$$

$$u_t + p_x = 0. \tag{4}$$

Our goal in this paper is to solve (3) and (4) using the Adomian decomposition method.

3. Adomian decomposition method

In this section, we present the Adomian decomposition method to solve the acoustic equations.

To see how the Adomian decomposition method works, let us start by noting derivative operators $L_t = \partial/\partial t$ and $L_x = \partial/\partial x$, so (3) and (4) become

$$L_t p + L_x u = 0, \tag{5}$$

$$L_t u + L_x p = 0. \tag{6}$$

The inverse for derivative operators for L_t and L_x are $L_t^{-1} = \int_0^t (.)dt$ and $L_x^{-1} = \int_0^t (.)dx$. In this paper, we only take the inverse with respect to the t variable. Applying the L_t^{-1} to the both sides of (5) and (6), we obtain

$$L_t^{-1}L_t p + L_t^{-1}L_x u = 0, \tag{7}$$

$$L_t^{-1}L_t u + L_t^{-1}L_x p = 0, \tag{8}$$

or

$$p(x,t) = p(x,0) - L_t^{-1}L_x u, \tag{9}$$

$$u(x,t) = u(x,0) - L_t^{-1}L_x p. \tag{10}$$

The variables $p(x,t)$ and $u(x,t)$ have to be written in series of Adomian polynomials

$$p(x,t) = \sum_{n=0}^{\infty} p_n(x,t), \tag{11}$$

$$u(x,t) = \sum_{n=0}^{\infty} u_n(x,t). \tag{12}$$

Applying the Adomian polynomials to the both sides, where $p_0 = p(x,0)$ and $u_0 = u(x,0)$, we obtain

$$\sum_{n=0}^{\infty} p_n = p(x,0) - L_t^{-1}L_x \sum_{n=0}^{\infty} u_n\,, \tag{13}$$

$$\sum_{n=0}^{\infty} u_n = u(x,0) - L_t^{-1}L_x \sum_{n=0}^{\infty} p_n\,, \tag{14}$$

or

$$p_0 + p_1 + p_2 + p_3 + \cdots = p(x,0) - L_t^{-1}\left(L_x\left(u_0 + u_1 + u_2 + u_3 + \cdots\right)\right), \tag{15}$$

$$u_0 + u_1 + u_2 + u_3 + \cdots = u\left(x,0\right) - L_t^{-1}\left(L_x\left(p_0 + p_1 + p_2 + p_3 + \cdots\right)\right). \tag{16}$$

The results of p and u decompositions are

$$p_0\left(x,t\right) = p\left(x,0\right), \tag{17}$$

$$p_1\left(x,t\right) = -L_t^{-1}\left(L_x\left(u_0\left(x,t\right)\right)\right), \tag{18}$$

$$p_2\left(x,t\right) = -L_t^{-1}\left(L_x\left(u_1\left(x,t\right)\right)\right), \tag{19}$$

$$p_3\left(x,t\right) = -L_t^{-1}\left(L_x\left(u_2\left(x,t\right)\right)\right), \tag{20}$$

$$\vdots$$

$$u_0\left(x,t\right) = u\left(x,0\right), \tag{21}$$

$$u_1\left(x,t\right) = -L_t^{-1}\left(L_x\left(p_0\left(x,t\right)\right)\right), \tag{22}$$

$$u_2\left(x,t\right) = -L_t^{-1}\left(L_x\left(p_1\left(x,t\right)\right)\right), \tag{23}$$

$$u_3\left(x,t\right) = -L_t^{-1}\left(L_x\left(p_2\left(x,t\right)\right)\right), \tag{24}$$

$$\vdots$$

For computational experiments in the next section, we choose initial conditions

$$p(x,0) = 0.1\,\mathrm{sech}^2(0.2x), \tag{25}$$

$$u\left(x,0\right) = 0, \tag{26}$$

for (3) and (4). We choose hyperbolic secant function because the function is smooth, so it has continuous derivatives. The amplitude phase and constant are taken to be 0.1 and 0.2, respectively. In all calculations in this paper, all quantities are assumed to have SI units with the MKS system.

4. Computational results

Adomian decomposition method needs some recursive iterations to get the approximation to the exact solution. We note that more iterations lead to more accurate solution with this method if the series is not yet convergent to the exact solution.

Using initial conditions (25) and (26), Adomian decomposition method leads to the following series formulas

$$p_{k+1}\left(x,t\right) = -L_t^{-1} L_x \sum_{k=0}^{\infty} u_k\,, \quad k \geq 0\,, \tag{27}$$

$$u_{k+1}\left(x,t\right) = -L_t^{-1} L_x \sum_{k=0}^{\infty} p_k\,, \quad k \geq 0\,, \tag{28}$$

where the exact solution is given by

$$\lim_{n\to\infty} P_n = p\left(x,t\right), \tag{29}$$

$$\lim_{n\to\infty} U_n = u\left(x,t\right). \tag{30}$$

The n-term approximations of the density p and the velocity u are

$$P_n[p] = \sum_{k=0}^{n-1} p_k\left(x,t\right), \quad n \geq 0, \tag{31}$$

$$U_n[u] = \sum_{k=0}^{n-1} u_k\,(x,t), \quad n \geq 0. \tag{32}$$

Using the Maple software, we obtain that the results of the iterations (up to p_4) for the pressure solution to our problem are expressed as follows:

$$p_0 = \frac{1}{10}\mathrm{sech}\left(\frac{1}{5}x\right)^2, \tag{33}$$

$$p_1 = 0, \tag{34}$$

$$p_2 = \frac{1}{125}\mathrm{sech}\left(\frac{1}{5}x\right)^2 \tanh\left(\frac{1}{5}x\right)^2 t^2 - \frac{1}{50}\mathrm{sech}\left(\frac{1}{5}x\right)^2 \left(\frac{1}{5} - \frac{1}{5}\tanh\left(\frac{1}{5}x\right)^2\right) t^2, \tag{35}$$

$$p_3 = 0, \tag{36}$$

$$p_4 = \frac{1}{9375}\mathrm{sech}\left(\frac{1}{5}x\right)^2 \tanh\left(\frac{1}{5}x\right)^4 t^4 - \frac{11}{3750}\mathrm{sech}\left(\frac{1}{5}x\right)^2 \tanh\left(\frac{1}{5}x\right)^2 \left(\frac{1}{5} - \frac{1}{5}\tanh\left(\frac{1}{5}x\right)^2\right) t^4$$
$$+ \frac{1}{375}\mathrm{sech}\left(\frac{1}{5}x\right)^2 \left(\frac{1}{5} - \frac{1}{5}\tanh\left(\frac{1}{5}x\right)^2\right)^2 t^4. \tag{37}$$

Furthermore, the results of the iterations (up to u_4) for the velocity solution to our problem are expressed as follows:

$$u_0 = 0, \tag{38}$$

$$u_1 = \frac{1}{25}\mathrm{sech}\left(\frac{1}{5}x\right)^2 \tanh\left(\frac{1}{5}x\right) t, \tag{39}$$

$$u_2 = 0, \tag{40}$$

$$u_3 = \frac{2}{1875}\mathrm{sech}\left(\frac{1}{5}x\right)^2 \tanh\left(\frac{1}{5}x\right)^3 t^3$$
$$- \frac{4}{375}\mathrm{sech}\left(\frac{1}{5}x\right)^2 \tanh\left(\frac{1}{5}x\right) \left(\frac{1}{5} - \frac{1}{5}\tanh\left(\frac{1}{5}x\right)^2\right) t^3, \tag{41}$$

$$u_4 = 0. \tag{42}$$

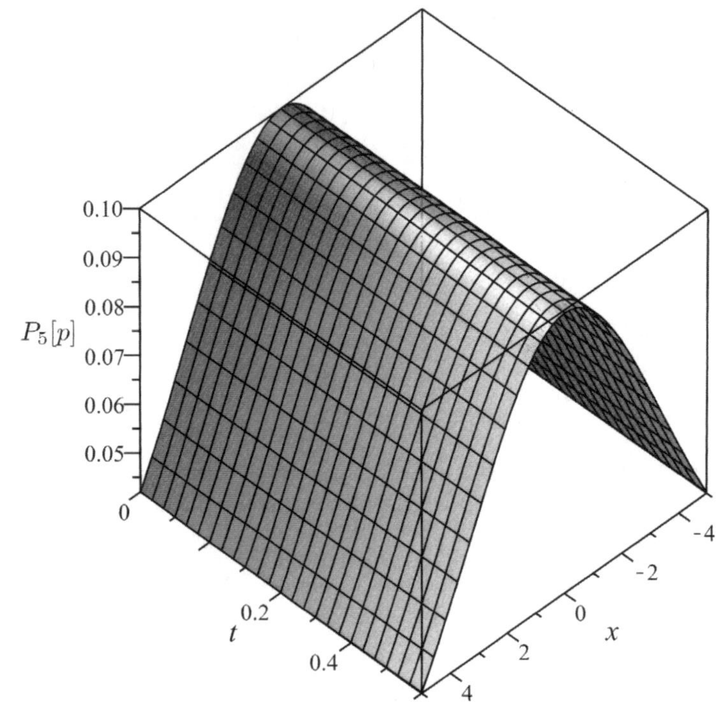

Figure 1. The approximate pressure $P_5[p]$ of the Adomian decomposition.

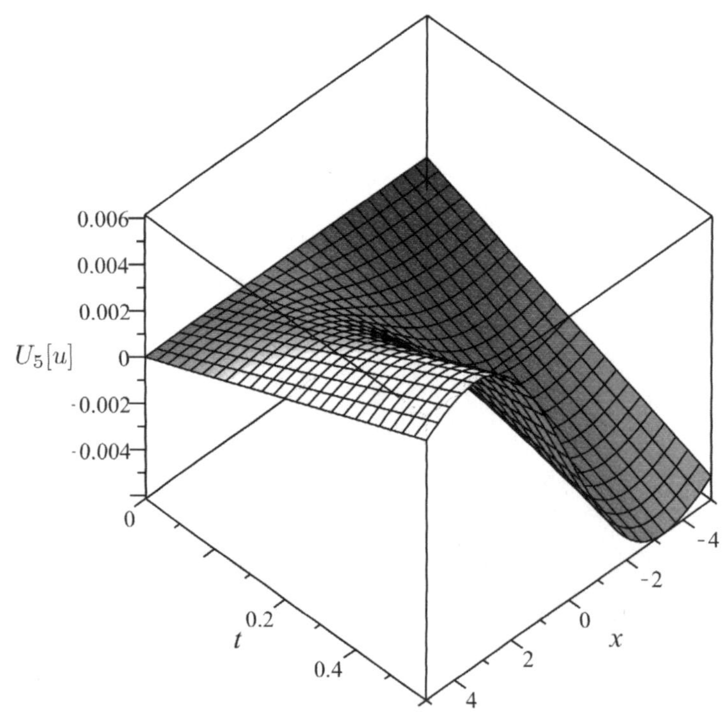

Figure 2. The approximate velocity $U_5[u]$ of the Adomian decomposition.

Now, we compute $P_5[p] = p_0 + p_1 + p_2 + p_3 + p_4$ and $U_5[u] = u_0 + u_1 + u_2 + u_3 + u_4$ using the above results. Therefore, we obtain that the approximate pressure and velocity are

$$\begin{aligned} P_5 = & \frac{1}{10}\operatorname{sech}\left(\frac{1}{5}x\right)^2 + \frac{1}{125}\operatorname{sech}\left(\frac{1}{5}x\right)^2 \tanh\left(\frac{1}{5}x\right)^2 t^2 \\ & - \frac{1}{50}\operatorname{sech}\left(\frac{1}{5}x\right)^2 \left(\frac{1}{5} - \frac{1}{5}\tanh\left(\frac{1}{5}x\right)^2\right) t^2 \\ & + \frac{1}{9375}\operatorname{sech}\left(\frac{1}{5}x\right)^2 \tanh\left(\frac{1}{5}x\right)^4 t^4 \\ & - \frac{11}{3750}\operatorname{sech}\left(\frac{1}{5}x\right)^2 \tanh\left(\frac{1}{5}x\right)^2 \left(\frac{1}{5} - \frac{1}{5}\tanh\left(\frac{1}{5}x\right)^2\right) t^4 \\ & + \frac{1}{375}\operatorname{sech}\left(\frac{1}{5}x\right)^2 \left(\frac{1}{5} - \frac{1}{5}\tanh\left(\frac{1}{5}x\right)^2\right)^2 t^4, \end{aligned} \tag{43}$$

and

$$\begin{aligned} U_5 = & \frac{1}{25}\operatorname{sech}\left(\frac{1}{5}x\right)^2 \tanh\left(\frac{1}{5}x\right) t + \frac{2}{1875}\operatorname{sech}\left(\frac{1}{5}x\right)^2 \tanh\left(\frac{1}{5}x\right)^3 t^3 \\ & - \frac{4}{375}\operatorname{sech}\left(\frac{1}{5}x\right)^2 \tanh\left(\frac{1}{5}x\right) \left(\frac{1}{5} - \frac{1}{5}\tanh\left(\frac{1}{5}x\right)^2\right) t^3, \end{aligned} \tag{44}$$

respectively.

Results of the pressure P_5 the velocity U_5 are plotted in figure 1 and figure 2, respectively. From these figures, the pressure from the origin point propagates to the left and to the right directions as the time evolves. The velocity mimics the pressure wave propagation, because the velocity is negative when the pressure wave propagates to the left, and the velocity is positive when the pressure wave propagates to the right. This is the correct behavior that we expect. In figure 2, the velocity tends to zero for large x and t.

5. Conclusion

We have solved the acoustic equations using the Adomian decomposition method. The method produces solutions any time and space point. The Adomian decomposition method can be implemented in a computer software with an inexpensive computation. The method is projected to be successful for solving multidimensional problems of acoustics.

Acknowledgments

The work in this paper was financially supported by Sanata Dharma University, which is gratefully acknowledged by the authors.

References

[1] LeVeque R J 2002 *Int. J. Numer. Methods Fluids* **40** 93
[2] Wazwaz A M 2009 *Partial Differential Equations and Solitary Waves Theory* (Beijing: Higher Education Press and Berlin: Springer)
[3] Adomian G 1998 *Appl. Math. Lett.* **11** 121
[4] Al-Khaled K and Allan F H 2004 *Math. Comput. Simul.* **66** 479

IOP Conf. Series: Journal of Physics: Conf. Series **856** (2017) 012003 doi:10.1088/1742-6596/856/1/012003

[5] Mungkasi S and Dheno M F S 2017 *AIP Conf. Proc.* **1788** 030103
[6] Dispini M and Mungkasi S 2016 *AIP Conf. Proc.* **1746** 020055
[7] Supriyadi B and Mungkasi S 2016 *Int. J. Multiscale Comput. Eng.* **14** 479
[8] Mungkasi S and Ningrum G I J 2016 *AIP Conf. Proc.* **1746** 020056

IOP Conf. Series: Journal of Physics: Conf. Series **856** (2017) 012004 doi:10.1088/1742-6596/856/1/012004

Equation of state within gluon dominated QGP model in relativistic hydrodynamics approach

T P Djun[1,2], M K N Patmawijaya[3], R Utama[3] and L T Handoko[2]

[1] Graduate Study in Materials Science, University of Indonesia, Kampus UI Salemba, Jakarta 10430, Indonesia

[2] Group for Theoretical and Computational Physics, Research Center for Physics, Indonesian Institute of Sciences, Kompleks Puspiptek Serpong, Tangerang 15310, Indonesia

[3] Physics Department, FMIPA, University of Indonesia, Depok 16424, Indonesia

E-mail: tp.djun@sci.ui.ac.id

Abstract. The dynamics of quark-gluon plasma (QGP) as a lump of deconfined free quarks and gluons is elaborated. Based on the first principal, we construct the Lagrangian that represents the dynamics of QGP. To induce a hydrodynamics approach, we substitute the gluon fields with flow fields. As a result, the derived equation of motion (E.O.M) for gluon dominated QGP shows a form similar to Euler equation; and the energy momentum tensor also represents explicitly the system of ideal fluid. Combining the E.O.M and energy momentum tensor, the pressure and energy density distribution are analytically derived.

1. Introduction

Recent experiments on heavy-ion collisions show a strong indication that hot dense deconfined phase of free quark and gluon, the so-called quark-gluon plasma (QGP), is conjectured to exist. The study of QGP itself has been carried out through a number of different approaches in previous works. Some results of these studies were obtained in the framework of quantum chromodynamics (QCD) theory by utilizing the lattice gauge calculation [1,2]. Other calculations of QGP were based on the relativistic hydrodynamics approach [3, 4]. In the latter, the QGP could be either quark- [4] or gluon- [3] dominated matter. For the quark-dominated approach, a very small ratio of shear viscosity over entropy is required to get a good fit of the spectra of transverse momentum, energy density distribution and other physical observables that are obtained from experiments [5–10]. On the other hand, the gluon-dominated plasma motivated by the discoveries of jet quenching in the heavy-ion collision at Brookhaven's Relativistic Heavy Ion Collider (RHIC) indicates the shock waves in the form of Mach cone [11,12]. The present paper adopts the so-called fluid QCD model [13–15] to produce the equation of motion and energy-momentum tensor for quark and gluon in a lump of QGP, and subsequently to investigate the distributions of the pressure and energy density.

This paper is organized as follows. In Section 2 the fluid QCD model is briefly revisited, and the energy momentum tensor for ideal fluid is derived. Then, it is followed by the derivation for the equation of state and the explicit expression of gluon field in Section 3. Finally, the summary and discussion will be given in Section 4. Throughout this work, we use the natural units, i.e., $\hbar = c = 1$.

IOP Conf. Series: Journal of Physics: Conf. Series **856** (2017) 012004 doi:10.1088/1742-6596/856/1/012004

2. Model

The Lagrangian of QGP that describes the unification of fermions and bosons from different gauge groups with preserving $\mathrm{SU}(3)_F \otimes \mathrm{U}(1)_G$ gauge symmetry can be written as [13, 14]

$$\mathcal{L} = i\bar{Q}\gamma^\mu \partial_\mu Q - m_Q \bar{Q}Q - \frac{1}{4}S^a_{\mu\nu}S^{a\mu\nu} - \frac{1}{4}F_{\mu\nu}F^{\mu\nu} + g_F J^a_{F\mu} U^{a\mu} + g_G J_{G\mu} A^\mu . \tag{1}$$

$\mathrm{SU}(3)_F \otimes \mathrm{U}(1)_G$ represents a non-Abelian $\mathrm{SU}(3)_F$ fluid fields containing quarks and antiquarks that interact with an electromagnetic field. Q and $\bar{Q}$ represent the quark and antiquark triplets, γ^μ are Dirac gamma matrices, and m_Q is the mass of quark. The term $\frac{1}{4}S^a_{\mu\nu}S^{a\mu\nu}$ is the gauge invariant kinetic term of gluon field. The gluon field strength tensor itself is expressed as $S^a_{\mu\nu} \equiv \partial_\mu U^a_\nu - \partial_\nu U^a_\mu + g_F f^{abc} U^b_\mu U^c_\nu$, where U^a_μ indicates gluon fields, g_F is the strong coupling constant, and f^{abc} is the structure constant of SU(3) gauge group. Similarly, the field strength tensor for A_μ is given by $F_{\mu\nu} = \partial_\mu A_\nu - \partial_\nu A_\mu$. The last two terms of (1), $J^a_{F\mu} = \bar{Q}T^a_F \gamma_\mu Q$ and $J_{G\mu} = \bar{Q}\gamma_\mu Q$, represent the quark currents from the SU(3) and U(1) gauge groups, respectively, whereas g_G denotes the coupling constant of U(1) gauge group.

The QCD Lagrangian in (1) is constructed with a purpose of reproducing the energy-momentum tensor that has the same form as the energy momentum tensor of an ideal fluid [13]. In the succeeding step, we propose that each gluon field U^a_μ, which is designed to act as the flow field in the system, be formulated in a configuration that inherently contains the relativistic flow. It is formulated as $U^a_\mu = (U^a_0, \mathbf{U}^a) \equiv u_\mu \phi^a$ [14, 15]. Here, ϕ^a is the dimension one scalar field, and $u_\mu \equiv \gamma(1, \mathbf{v})$ is the relativistic velocity, with $\gamma = (1 - |\mathbf{v}|^2)^{-1/2}$.

This formulation provides a clue that a single gluonic field U^a_μ may behave as a fluid at certain scale beside its conventional point particle properties with a polarization vector ϵ_μ in the form of $U^a_\mu = \epsilon_\mu \phi^a$. One can, then, consider that there is a kind of "phase transition",

$$\underbrace{\text{hadronic state}}_{\epsilon_\mu} \longleftrightarrow \underbrace{\text{QGP state}}_{u_\mu} .$$

As the gluon field behaves as a point particle, it is in a stable hadronic state and is characterized by its polarization vector. On the other hand in the pre-hadronic state (before hadronization) like hot QGP, the gluon field behaves as a highly energized flow particle, and the properties are dominated by its relativistic velocity.

The field U^a_μ is actually analogous to the gauge boson from $U(1)$ gauge group in particle physics, where the polarization vector ϵ_μ from the free photon solution $\sim \epsilon_\mu \exp(-ip_\mu x^\mu)$ is replaced by the 4-velocity u_μ. Recall that the wavefunction U^a_μ for a free particle satisfies $[g^{\mu\nu}(\partial^2 + m^2) - \partial^\nu \partial^\mu]U^a_\mu = 0$ with solution $U^a_\mu \sim \epsilon_\mu \, \phi^a \exp(-ip_\mu x^\mu)$. For a massive vector particle, $m \neq 0$, so we have no choice but taking $\partial^\mu U^a_\mu = 0$. It is not a gauge condition like the case of massless particle. This then demands $p^\mu \epsilon_\mu = 0$. The number of independent polarization vectors is reduced from four to three in a covariant fashion. However, one can still perform another gauge transformation to the massless U^a_μ, which makes finally only two degrees of freedom. Therefore, one should keep in mind that in the present model the spatial velocity has only two degrees of freedom, which means that one component must be described by another two vector components.

Furthermore, when Euler-Lagrange equation is applied to (1), one obtains

$$\frac{\partial}{\partial t}(\gamma \mathbf{v} \phi^a) + \nabla(\gamma \phi^a) = -g_F \oint d\mathbf{x}(J^a_{F0} + F^a_0) , \tag{2}$$

where J^a_{F0} denotes the covariant current originating from the quarks that are surrounded by and interact with the gluon "fluid", while the term F^a_μ is induced by the fluid self-interaction and the interacting gauge fields A^a_μ. Equation (2) is considered as the general relativistic fluid equation for single gluon field U^a_μ since in the nonrelativistic limit, i.e. $\phi^a \sim 1$ and $\gamma \sim 1 + \frac{1}{2}|\mathbf{v}|^2$, such equation transforms to the classical equation of motion of fluid dynamics [14]

$$\frac{\partial \mathbf{v}}{\partial t} + (\mathbf{v} \cdot \nabla)\mathbf{v} = -g_F \oint d\mathbf{x}(J^a_{F0} + F^a_0) \mid_{\text{non-rel}} . \tag{3}$$

This shows that from a certain point of view the Lagrangian describes a general relativistic fluid system interacting with another gauge field and the matter inside. The flow characteristic that appears in the equation of motion comes from the contribution of gluon field U^a_μ. This fact indicates that the system we are working with is a gluon-dominated QGP. In such system the terms that do not contain gluon field may be omitted. As a consequence, we have

$$\mathcal{L} = -\tfrac{1}{4} S^a_{\mu\nu} S^{a\mu\nu} + g_F J^a_{F\mu} U^{a\mu} . \tag{4}$$

The Lagrangian given by (4) describes the kinetics of gluons, the self-interactions of gluons, and the interaction between quark with gluon "fluid". Electromagnetic interactions also exist in the system, but they are suppressed due to the tiny value of electromagnetic coupling compared with that of the strong interaction. From the Lagrangian given in (4), we can derive the energy momentum tensor [16]

$$\begin{aligned}
\mathcal{T}_{\mu\nu} &= \frac{2}{\sqrt{-g}} \frac{\delta(\mathcal{L}\sqrt{-g})}{\delta g^{\mu\nu}} \\
&= S^a_{\mu\rho} S^{a\rho}{}_\nu - g_{\mu\nu}\mathcal{L} + 2 g_F J^a_{F\mu} U^a_\mu \\
&= [2 g_F g_{\mu\nu} J^a_{F\mu} U^{a\mu} + g_F^2 f^{abc} f^{ade} U^b_\mu U^c_\rho U^{d\rho} U^e_\nu] \\
&\quad -[g_F g_{\mu\nu} J^a_{F\mu} U^{a\mu} - \frac{1}{4} g_{\mu\nu} g_F^2 f^{abc} f^{ade} U^b_\mu U^c_\nu U^{d\mu} U^{e\nu}] .
\end{aligned} \tag{5}$$

The factor $J^a_{F\mu} U^{a\mu}$ in (5) might be expressed in a more elementary form. To this end, we can start with the assumption that the solution of Dirac equation for single color quark is in the form of $Q(p, x) = q(p)\exp(-ip \cdot x)$. By inserting this solution to the Dirac equation, i.e., $(i\gamma^\mu \partial_\mu - m)q(p)\exp(-ip \cdot x) = 0$, and multiplying with the antiquark solution, we obtain $\bar{q}\gamma_\mu q = 4p_\mu$. Since $J^a_{F\mu} = \bar{Q}\gamma_\mu Q T^a = \bar{q}\gamma_\mu q T^a = 4p_\mu T^a$, and $U^{a\mu} = u^\mu \phi^a$, we finally obtain $J^a_{F\mu} U^{a\mu} = 4m_Q T^a \phi^a$. Note that we have utilized $u_\mu u^\mu = 1$ along with the assumption that all quarks and antiquarks have the same momenta P^μ, and that the velocities of all gluons and quarks are homogeneous. Furthermore, due to the completely anticommutative property of the structure constant f^{abc}, the 2nd and 4th terms in (5) vanish. Then, the energy-momentum tensor in the function of field ϕ^a reads

$$\mathcal{T}_{\mu\nu} = [8 g_F m_Q T^a \phi^a] u_\mu u_\nu - [4 g_F m_Q T^a \phi^a] g_{\mu\nu} . \tag{6}$$

With this form, $\mathcal{T}_{\mu\nu}$ is obviously describing a system of perfect fluid. Furthermore, we assume that the gluon color states are homogeneous, i.e., $\phi^a = \phi$ for all $a = 1, \cdots, 8$. As the follow-on to this assumption, the generator T^a can be compactly written as $T = \sum_a T^a$. The energy-momentum tensor then reduces to

$$\mathcal{T}_{\mu\nu} = (8T g_F m_Q \phi) u_\mu u_\nu - (4T g_F m_Q \phi) g_{\mu\nu}. \tag{7}$$

It reveals the collective gluons flow in the system. Therefore, we can temporarily summarize that the system represents an isotropic homogeneous perfect fluid of gluon-dominated QGP. Furthermore, to discuss the dynamics of glue lumps, it is also plausible to take the gluon as a field that is independent of time, $\phi = \phi(x)$. If we compare (7) with energy-momentum tensor of ideal fluid $\mathcal{T}_{\mu\nu} = (\mathcal{E} + \mathcal{P}) u_\mu u_\nu$ - $\mathcal{P} g_{\mu\nu}$, it becomes obvious that

$$\mathcal{E} = \mathcal{P} = 4T g_F m_Q \phi. \tag{8}$$

Here $\mathcal{P}$ and $\mathcal{E}$ denote the isotropic pressure and energy density for fluid field, respectively. Or, in currently discussed topic, it is the energy density and pressure distributions in the lump of gluon dominated QGP.

3. Expression for distribution field

The expression for the scalar field ϕ can be obtained by deriving the solution of (3). After applying the assumption of the gluon homogeneity, the equation can be expressed as follows.

$$\delta\partial_{tr}\phi + \gamma\partial_{rr}\phi + \xi + \beta\phi\partial_t\phi + \beta\phi\partial_r\phi - \lambda\partial_t\phi - \lambda\partial_r\phi = 0, \tag{9}$$

with $\delta = \gamma|\mathbf{v}|$, $\xi = g_F\bar{Q}\gamma_\mu TQ$, $\beta = i\gamma^2|\mathbf{v}|^2$, $\lambda = (g_G/g_F)A\gamma$, and $\gamma = (1-v^2)^{-\frac{1}{2}}$. To simplify but stay relevant, the pressure and energy density distribution of the lump of QGP is assumed to depend only on r, that is, $\phi = \phi(r)$ and $\partial_r \to d_r$. Therefore, the terms that involve the derivative of t vanish,

$$\gamma d_{rr}\phi + \beta\phi d_r\phi - \lambda d_r\phi + \xi = 0. \tag{10}$$

The solution for this equation is

$$\begin{aligned}\phi &= \Big(2\gamma\Big(\frac{\exp[\frac{r\lambda}{2\gamma}]\Lambda_1[x]}{2\gamma} - \frac{\exp[\frac{r\lambda}{2\gamma}]\beta\xi\Lambda_2[x]}{2^{1/3}\gamma^2(-\frac{\beta\xi}{\gamma^2})^{2/3}} \\ &\quad +\Big(\frac{\exp[\frac{r\lambda}{2\gamma}]\Lambda_3[x]}{2\gamma} - \frac{\exp[\frac{r\lambda}{2\gamma}]\beta\xi\Lambda_4[x]}{2^{1/3}\gamma^2(-\frac{\beta\xi}{\gamma^2})^{2/3}}\Big)C_2\Big)\Big) \\ &\quad \times 1/\Big(\beta\Big(\exp[\frac{r\lambda}{2\gamma}]\Lambda_1[x] + \exp[\frac{r\lambda}{2\gamma}]\Lambda_3[x]C_2\Big)\Big).\end{aligned} \tag{11}$$

When we substitute $\delta, \xi, \beta, \lambda$ and γ back to (11), then ϕ appears as

$$\begin{aligned}\phi &= \Big(\frac{2}{\sqrt{1-v^2}}\Big(\frac{\exp[\frac{rgA}{2}]\Lambda_1[x]}{2/\sqrt{1-v^2}} - \frac{\exp[\frac{rgA}{2}](iv^2\xi)^{1/3}\Lambda_2[x]}{2^{1/3}} \\ &\quad +\Big(\frac{\exp[\frac{rgA}{2}]\Lambda_3[x]}{2/\sqrt{1-v^2}} - \frac{\exp[\frac{rgA}{2}](iv^2\xi)^{1/3}\Lambda_4[x]}{2^{1/3}}\Big)C_2\Big)\Big) \\ &\quad \times 1/\Big(\frac{iv^2}{1-v^2}\Big(\exp[\frac{rgA}{2}]\Lambda_1[x] + \exp[\frac{rgA}{2}]\Lambda_3[x]C_2\Big)\Big).\end{aligned} \tag{12}$$

Here,

$$\begin{aligned}\Lambda_1[x] &= \frac{1}{3^{1/6}\Gamma[\frac{2}{3}]} + \frac{3^{1/6}x}{\Gamma[\frac{1}{3}]} + \frac{x^3}{6(3^{1/6}\Gamma[\frac{2}{3}])} + \frac{x^4}{4(3^{5/6}\Gamma[\frac{1}{3}])} + \mathcal{O}[x]^5\ , \\ \Lambda_2[x] &= \frac{3^{1/6}}{\Gamma[\frac{1}{3}]} + \frac{x^2}{2(3^{1/6}\Gamma[\frac{2}{3}])} + \frac{x^3}{3^{5/6}\Gamma[\frac{1}{3}]} + \frac{x^5}{30(3^{1/6}\Gamma[\frac{2}{3}])} + \mathcal{O}[x]^6\ , \\ \Lambda_3[x] &= \frac{1}{3^{2/3}\Gamma[\frac{2}{3}]} - \frac{x}{3^{1/3}\Gamma[\frac{1}{3}]} + \frac{x^3}{6(3^{2/3}\Gamma[\frac{2}{3}])} - \frac{x^4}{12(3^{2/3}\Gamma[\frac{2}{3}])} + \mathcal{O}[x]^5\ , \\ \Lambda_4[x] &= -\frac{1}{3^{1/3}\Gamma[\frac{1}{3}]} + \frac{x^2}{2(3^{2/3}\Gamma[\frac{2}{3}])} - \frac{x^3}{3(3^{1/3}\Gamma[\frac{1}{3}])} + \frac{x^5}{30(3^{2/3}\Gamma[\frac{2}{3}])} + \mathcal{O}[x]^6\ ,\end{aligned} \tag{13}$$

IOP Conf. Series: Journal of Physics: Conf. Series **856** (2017) 012004 doi:10.1088/1742-6596/856/1/012004

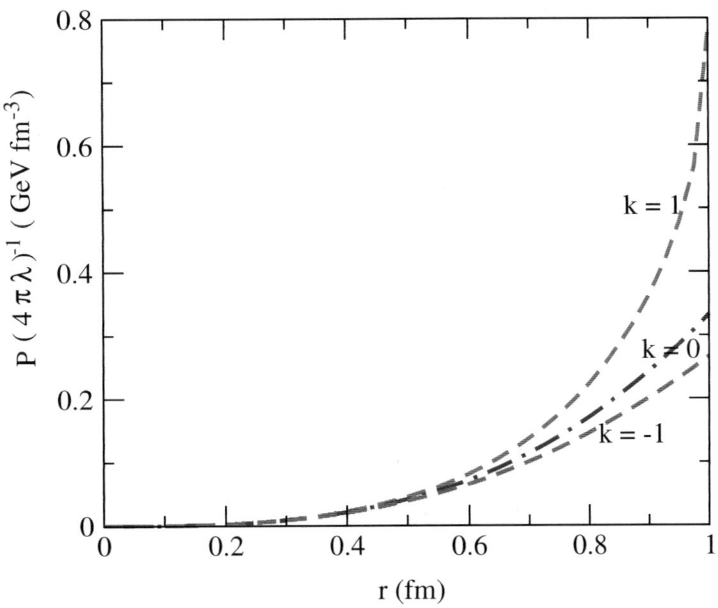

Figure 1. Pressure inside QGP-lump in the function of radius.

and

$$x = \frac{-i2v^2(\xi r + \frac{C_1}{\sqrt{1-v^2}}) + g^2 A^2}{2^{4/3}(-iv^2\xi)^{2/3}}. \tag{14}$$

So far, this nontrivial solution is well expressed. But to have a firm solution, some adjustments still need to be carried out in future works. For the rest, we will explore the expression for pressure and energy density in the system of gluon-dominated quark-gluon plasma.

4. Summary and discussion

We return to (8), $\mathcal{E} = \mathcal{P} = 4Tg_F m_Q \phi$, the energy and pressure can be expressed as follows [17]

$$\rho = P = \int \mathcal{E} d^4x = \int \mathcal{P} d^4x = 4Tg_F f_Q m_Q \int\int \phi dt dV \ , \tag{15}$$

where $\mathcal{E}$ and $\mathcal{P}$ denote the isotropic pressure and density for single fluid field, respectively.

By adopting FRW geometry as the background, and doing integration on the spatial dimension, we get

$$\rho = P = 4\pi\lambda \int R^3 \phi dt \int \frac{r^2 dr}{\sqrt{1-kr^2}}. \tag{16}$$

Here we have used $dV = (R^3 r^2 \sin\theta)/\sqrt{1-kr^2} dr d\theta d\vartheta$ and $\lambda = 4Tg_F f_Q m_Q$. Furthermore, if we assign $\int_0^{2\pi}\int_0^{\pi}\int dV = 4\pi \int \frac{r^2 dr}{\sqrt{1-kr^2}}$ as ζ, then the expression for P and ρ can be written as

$$P = \rho = \lambda\zeta \int R^3 \phi dt. \tag{17}$$

In a case when r is constant, $p = \rho \propto \phi(t)dt$; and also if $\phi(t) \propto 1/t$, then one arrives at $P = \rho \propto \lambda\zeta R^3/t$. It indicates that the pressure and energy in the system decrease asymptotically following the increases of time.

Figure 1 shows the total pressure inside the QGP-lump as a function of radius. It is drawn for the simplest condition, where ϕ is assumed as a constant, and the scale factor $R = 1$. $k = 1, 0, -1$ represent the space-time curvatures that describe the close, flat and open space-time. One of the obvious properties that appears here is the equation of state $P/\rho = 1$. Such a ratio indicates that in this model the QGP exists within radiation state. This estimate comes from the fact that in general the prerequisite condition for the radiation state is $P \sim \rho/3$, and P/ρ is getting smaller along the transition from radiation state to matter state.

Finally, the study reveals observables that should be accessible through heavy ion collisions experiments at the RHIC and Large Hadron Collider (LHC). With currently proposed calculation results, such experiments are in future expected to become references for adjustments and verifications of the dynamics theory of macroscopic behavior of QGP.

Acknowledgments
T.P.D thanks the Group for Theoretical and Computational Physics, Research Center for Physics, Indonesian Institute of Sciences (LIPI) for the warm hospitality during the completion of this work. T.P.D. and L.T.H. are supported by Riset Kompetitif LIPI under Contract No. 11.04/SK/KPPI/II/2016.

References
[1] Gottlieb S 2007 *J. Phys.: Conf. Series* **78** 012023
[2] Petreczky P 2008 *Eur. Phys. J.* ST **155** 123
[3] Bouras I, Molnar E, Niemi H, Xu Z, El A, Fochler O, Greiner C and Rischke D H 2009 *Phys. Rev. Lett.* **103** 032301
[4] Romatschke P 2010 *Int. J. Mod. Phys.* E **19** 1
[5] Teaney D, Lauret J and Shuryak E V 2001 *Phys. Rev. Lett.* **86** 4783
[6] Huovinen P, Kolb P F, Heinz U W, Ruuskanen P V and Voloshin S A 2001 *Phys. Lett.* B **503** 58
[7] Kolb P F, Heinz U W, Huovinen P, Eskola K J and Tuominen K 2001 *Nucl. Phys.* A **696** 197
[8] Kolb P F and Rapp R 2003 *Phys. Rev.* C **67** 044903
[9] Hirano T and Tsuda K 2002 *Phys. Rev.* C **66** 054905
[10] Baier R and Romatschke P 2007 *Eur. Phys. J.* C **51** 677
[11] Adams J *et al* 2003 *Phys. Rev. Lett.* **91** 172302
[12] Adare A *et al* 2008 *Phys. Rev. Lett.* **101** 232301
[13] Djun T P, Handoko L T, Soegijono B and Mart T 2015 *Int. J. Mod. Phys.* A **30** 1550077
[14] Sulaiman A, Fajarudin A, Djun T P and Handoko L T 2009 *Int. J. Mod. Phys.* A **24** 3630
[15] Djun T P and Handoko L T 2010 *Proc. Conf. in Honour of Murray Gell-Mann's 80th Birthday: Quantum Mechanics, Elementary Particles, Quantum Cosmology and Complexity* (Singapore: Nanyang Technological University) p 419
[16] Hobson M P, Efstathiou G and Lasenby A N 2006 *General Relativity: An Introduction for Physicists* (Cambridge: University Press)
[17] Nugroho C S, Latief A O, Djun T P and Handoko L T 2012 *Grav. Cosmol.* **18** 32

Nonlinear dynamics of DNA bubble induced by site specific DNA-protein interaction

Donny Dwiputra[1], **Wahyu Hidayat**[1,2] **and Freddy P Zen**[1,2]

[1] Theoretical Physics Laboratory, Faculty of Mathematics and Natural Sciences, Institut Teknologi Bandung, Jl. Ganesha no. 10, Bandung 40132, Indonesia
[2] Indonesian Center of Theoretical and Mathematical Physics (ICTMP), Jl. Ganesha no. 10, Bandung 40132, Indonesia

E-mail: donny.dwiputra@s.itb.ac.id

Abstract. We propose a dynamical model describing interactions between DNA and a specific binding protein involving long-range transmission of biological information. The model couples the hydrogen bonds between the one connecting DNA and protein side chain and the one connecting DNA base pairs since they account for site specificity of the binding. We adopt Morse potentials with coupling terms to construct model of coupled hydrogen bonds. We show that the model gives rise to a breather soliton formation, corresponding to the DNA bubbles, which propagates through DNA chain as the carrier of genetic information. We investigate various kinds of possible coupling dynamics and suggest the model realism in depicting the renaturation or hybridization processes.

1. Introduction

Lossless communication between DNA-binding proteins is the basis of life because it facilitates the gene regulation, transcription and replication of DNA. It has been proposed [1–4] that a protein induces propagating DNA deformations that play role as the carrier for lossless biological communication in DNA chain [5]. Induced conformation in DNA is referred as allosteric effect in DNA. Nonlinearity is necessary since any harmonic wave could not possess a coherent wave in dispersive medium and highly dissipative environment. Thus, the propagating conformation is nicely depicted by a solitonic wave. In this paper, we aim to investigate the dynamics of solitonic wave corresponding to the propagating local conformational distortion in DNA that acts as the information carrier induced by the binding of protein in specific DNA sites. A discussion in biological terms will be conducted to give the mathematical solution a meaning. The relevant preprint is also available in [6].

The notable study of nonlinear localized soliton that propagates through DNA was conducted by Peyrard and Bishop (PB) [7], with their notable breather excitation referred as DNA bubble, which has experimentally been measured in [8]. The dynamics and thermal effects of PB breather are studied in [9,10], while the effect of viscosity and external forces are investigated in [11]. The PB model has been modified to include the protein interaction, such as the statistical model given in [12] and TFAM (Transcription Factor A, Mitochondrial)-DNA interaction model [13] that depicts the allosteric interactions by proteins. However, the effect of the chemical bonds between protein and DNA to an analytical breather is not yet investigated.

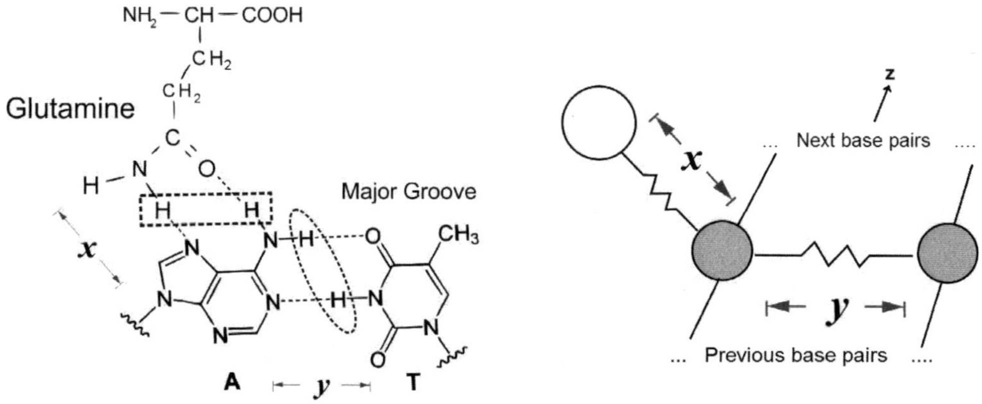

Figure 1. Protein-DNA H-bonding. We pick the example case glutamine chain binding to Adenine (**A**)-Thymine (**T**) base pair. The H-bonds are in box (x) and in oval (y).

The notable property of DNA-protein interaction is that the protein can bind to a specific region in DNA, depending on its certain DNA sequence or base-pair type. Origin of the specificity in DNA recognition by protein, at chemical signatures, is carried by the base pairs and the DNA shapes [14–16]. The chemical bonds involved are electrostatic, van der Waals's and hydrogen bond (H-bond); but the latter is the most significant because it possesses the highest transition free energy the most specific binding and the nonspecific case [17]. Thus, the H-bond is responsible for specific DNA-protein interaction.

In this paper, we will derive the solution to our DNA-protein model [18] by using multiple scale perturbation method and later investigate its dynamics. The model Hamiltonian is

$$H = \sum_n \left\{ \frac{p_{\mathrm{y}n}^2}{2m} + D_{(z)}(\mathrm{e}^{-\alpha y_n} - 1)^2 + \frac{k}{2}(y_n - y_{n-1})^2 + \frac{p_{\mathrm{x}m}^2}{2M} + E(\mathrm{e}^{-\beta x_m} - 1)^2 + \frac{\chi}{2} x_m^a y_n^b f_{mn} \right\}. \tag{1}$$

Here p_{yn} and y_n are momentum and position of the stretching mode of a base pair, respectively; while D is the depth of the Morse potential (second term) of this mode. The DNA-protein stretching x_m is controlled under a Morse potential by depth E and is interacting with DNA at site-m. The first three terms are DNA parts consisting of Morse potential for the base pair (the oval dots in figure 1) and the harmonic stacking interaction; the next two terms are protein parts, which also consist of a Morse potential for H-bonds connecting DNA and protein side chain (the rectangle dots in figure 1). The last term is interaction term with coupling constant χ; it contains Gaussian decay factor f_{mn} with σ^2 proportional to variance:

$$f_{mn} = \exp[-\sigma^2(m-n)^2]. \tag{2}$$

The dynamics of the DNA-protein interaction relies heavily on the values of a and b in the last term of (1). One can check the implication via perturbation method by implying $a = b = 1$. It will be seen that the interaction term pops up in the zeroth-order equations, and it is undesirable for solving perturbations [18]. By considering the biological facts that the protein triggers a local opening in DNA, even when all base pairs are still closed ($y_n = 0$ initially), it is required that the interaction force in the equation of motion for y_n not contain the amplitude y_n; otherwise, a closed DNA segment will forever be unopened even though the protein is interacting nearby. Here we take $a = 2$ and $b = 1$. The structural stability analysis of this Hamiltonian model is investigated in [19].

2. Nonlinear excitations in the DNA-protein interaction

2.1. The multiscale expansion method

To examine how the protein induces a local base-pair conformation (i.e., the opening), we must obtain the analytical form of the nonlinear excitation. A well-known perturbative method has been developed in [20], which is actually based on the multiscale expansion technique. The time t is expanded to $t_0, t_1, \ldots$ where $t_n = \epsilon^n t$ [21] and ϵ is a small parameter. By using multiple scales method, we treat fastly and slowly varying time and spatial scales separately so that we obtain envelope amplitudes in different scales. A naive perturbation cannot, in general, achieve such solutions.

According to original PB approach [22], it is assumed that the oscillations of bases be large enough to be anharmonic but still insufficient to break the H-bond since the Morse plateau is not yet reached. The shifts $Y_n \equiv \alpha y_n$ and $X_m \equiv \beta x_m$ oscillate around the bottom of symmetric potential; hence the transformations $Y_n = \epsilon\phi_n$ and $X_m = \epsilon\psi_m$ can be safely implemented to scale the equations of motion. Here we use the continuum approximation, assuming a long DNA chain with lattice space $a \to 0$, implying $na \to z$, $m/a \to \rho$, $ka \to K$, $D_{(z)}/a \to \mathcal{D}$, $f_{ml} \to f(z) = \exp[-\sigma^2(z-z_0)^2]$ and $\chi/a \to \mathcal{X}$. The continuum approximation brings us to the equations of motion for (1),

$$\phi_{tt} - S\phi_{zz} + V_{(z)}\phi = V_{(z)}\left(\frac{3}{2}\epsilon\phi^2 - \frac{7}{6}\epsilon^2\phi^3 + \mathcal{O}(\epsilon^3)\right) - \frac{\mu}{2}\epsilon\psi^2 f(z), \tag{3}$$

$$\psi_{tt} + W\psi = W\left(\frac{3}{2}\epsilon\psi^2 - \frac{7}{6}\epsilon^2\psi^3 + \mathcal{O}(\epsilon^3)\right) - \eta\epsilon\psi\int\phi f(z)\mathrm{d}z, \tag{4}$$

where the continuum parameters are

$$V_{(z)} = \frac{2\alpha^2\mathcal{D}}{\rho},\ W = \frac{2\beta^2 E}{M},\ S = \frac{K}{\rho},\ \mu = \frac{\mathcal{X}\alpha}{\rho\beta^2},\ \eta = \frac{\mathcal{X}}{M\alpha}. \tag{5}$$

For simplicity hereafter, we write $z \equiv z_0$, $Z \equiv z_1$, $t \equiv t_0$, $T \equiv t_1$, and $\tau \equiv t_2$. The multiscale expansion yields the time or spatial derivative for each order of ϵ^n so that we can solve the equations recursively. It should be borne in mind that $\phi = \phi(z, Z, t, T, \tau)$ and $\psi = \psi(t, T, \tau)$ [18]. This method breaks down the nonlinear problem into several nonhomogenous linear ordinary differential equations up to $\mathcal{O}(\epsilon^2)$ in ϕ and ψ; the solutions are [18]

$$\phi^{(0)} = A_1(Z, T, \tau)\mathrm{e}^{i\theta} + \text{c.c.}, \tag{6}$$

$$\psi^{(0)} = 2(\tau)\mathrm{e}^{i\varphi} + \text{c.c.}, \tag{7}$$

$$\phi^{(1)} = 3|A_1|^2 - \frac{\mu f(z)}{2\sigma V_{(z)}}|A_2|^2 - \frac{1}{2}A_1^2\mathrm{e}^{2i\theta} + \frac{\mu f(z)}{12\sigma V_{(z)}}A_2^2\mathrm{e}^{2i\varphi} + \text{c.c.},$$

$$\psi^{(1)} = 3|A_2|^2 - \frac{1}{2}A_2^2\mathrm{e}^{2i\varphi} + \frac{\eta\sqrt{\pi}}{\sigma}\left[\frac{A_2\int A_1\mathrm{d}Z}{\omega^2 + 2\omega\sqrt{W}}\mathrm{e}^{i(\tilde{\theta}+\varphi)} + \frac{A_2^*\int A_1\mathrm{d}Z}{\omega^2 - 2\omega\sqrt{W}}\mathrm{e}^{i(\tilde{\theta}-\varphi)}\right]\mathrm{e}^{-q^2/4\sigma^2} + \text{c.c.}, \tag{8}$$

where $\theta = qz - \omega t$, $\tilde{\theta} = qz_0 - \omega t$ and the phase $\varphi = \sqrt{W}t$. From here we get the dispersion relation,

$$\omega^2 = V_{(z)} + Sq^2. \tag{9}$$

Notice that $\partial A_2/\partial T = 0$, so A_2 has no dependence of T. Finally, the slow varying envelopes A_1 and A_2 are determined from the coupled nonlinear Schrödinger-like equations obtained from zeroing the secular terms ($\exp(\pm i\theta)$ and $\exp(\pm i\varphi)$) in $\mathcal{O}(\epsilon^2)$,

$$i\frac{\partial A_1}{\partial \tau} + P_1\frac{\partial^2 A_1}{\partial \xi^2} + Q_1|A_1|^2 A_1 = 3\mu f|A_2|^2 A_1, \tag{10}$$

$$i\frac{\partial A_2}{\partial \tau} + Q_2|A_1|^2 A_2 = \eta\gamma \int |A_1|^2 A_2 \, \mathrm{d}Z, \tag{11}$$

where

$$\gamma = \left[3 + \frac{2\eta\sqrt{\pi}}{\omega^2 - 4W}\mathrm{e}^{-q^2/4\sigma^2}\right]\frac{\sqrt{\pi}}{\sigma}, \quad Q_1 = 4V_{(z)},$$

$$Q_2 = \left[4W + \frac{5\mu\eta\sqrt{\pi/2}}{12\sigma^2 V_{(z)}}\right], \quad P_1 = \frac{S - V_g^2}{2\omega}, \tag{12}$$

and $\xi = Z - V_g T$ is a right-moving coordinate having group velocity $V_g = Sq/\omega$. The integral of $|A_1|^2$ with respect to Z over entire space is a finite function of time because we assume a localized solitonic wave.

2.2. Nonlinear excitations

The bright soliton solutions of coupled NLS have been fairly investigated, such as in [23]. We use the Hirota bilinear method [24] to solve (10) and (11) with the transformations

$$A_1 \equiv \frac{G(\xi,\tau)}{F(\xi,\tau)} \quad \text{and} \quad A_2 \equiv \left.\frac{H(\xi,\tau)}{F(\xi,\tau)}\right|_{\xi=\xi_0} \tag{13}$$

where $F \in \mathbb{R}$ and $G, H \in \mathbb{C}$. Here we apply a technique that assumes spatial dependency of A_2 in the first place, and discard it by inserting a constant $\xi = \xi_0$ after the solution is found. By inserting (13) into (10) and (11), we get the bilinear forms

$$\left(iD_\tau + P_1 D_\xi^2\right) G \cdot F = 0,$$

$$Q_1|G|^2 - \mu\gamma f|H|^2 = P_1 D_\xi^2 F \cdot F, \tag{14}$$

$$iD_\tau H \cdot F = 0,$$

$$Q_2|H|^2 = \eta\gamma|G|^2. \tag{15}$$

From (15) we can relate A_1 and A_2 by $|H|^2 = \eta\gamma|G|^2/Q_2$. The problem of finding one-soliton solution with a common phase is equivalent to solving the NLS in the form of

$$i\frac{\partial A_1}{\partial \tau} + P_1\frac{\partial^2 A_1}{\partial \xi^2} + Q'(z)|A_1|^2 A_1 = 0, \tag{16}$$

where

$$Q'(z) = Q_1 - \frac{3\mu\eta\gamma}{Q_2} f(z) \tag{17}$$

is obtained from (15). The solution for bright soliton, $PQ' > 0$, is [20]

$$A_1(\xi,\tau) = A\,\mathrm{sech}\left[A\left(\frac{Q'}{2P_1}\right)^{1/2}\left(\xi - \frac{v_e \tau}{P_1}\right)\right] \times \exp\left[i\left(\frac{v_e}{2P_1}\right)\left(\xi - \frac{v_c \tau}{P_1}\right)\right], \tag{18}$$

where A is the amplitude,

$$A(z) = \left(\frac{v_e^2 - 2v_e v_c}{2P_1 Q'(z)}\right)^{1/2}, \tag{19}$$

with condition $v_e^2 - 2v_e v_c > 0$. Both v_c and v_e are carrier and envelope wave velocities, respectively, where $v_c = g v_e$ for positive g. By taking a common phase, we get that the expression for A_2 just differs by the amplitude according to (15),

$$A_2(\tau) = \left(\frac{\eta\gamma}{Q_2}\right)^{1/2} A_1(\xi,\tau)|_{\xi=\xi_0, z=z_0}. \tag{20}$$

To obtain the solution, we first calculate the integral term in (8) by changing the domain Z to ξ,

$$\int_{-\infty}^{\infty} A_1\,\mathrm{d}Z = CA\exp\left[\frac{iv_e^2}{2P_1^2}(1-g)\tau\right], \tag{21}$$

where

$$C = \frac{2\pi P_1}{v_e\sqrt{1-2g}}\,\mathrm{sech}\left(\frac{\pi}{2\sqrt{1-2g}}\right). \tag{22}$$

Inserting (18) and (20) into (6)–(8) and setting ξ_0 and z_0 to zero, we get

$$\begin{aligned} Y(z,t) &= \epsilon 2A\,\mathrm{sech}\,\Theta\cos(N_+ t) + \epsilon^2 A^2\,\mathrm{sech}^2\Theta \\ &\quad\times\{3 - \cos 2(Qz - Mt) - \Lambda(z) \\ &\quad\times[1 - \cos 2((Q-q)z - Mt)]\} + \mathcal{O}(\epsilon^3), \end{aligned} \tag{23}$$

$$\begin{aligned} X(t) &= \epsilon 2A\left(\frac{\eta\gamma}{Q_2}\right)^{1/2}\mathrm{sech}\,\Theta_0\cos(N_+ t) + \epsilon^2 A^2\left(\frac{\eta\gamma}{Q_2}\right) \\ &\quad\times\mathrm{sech}^2\Theta_0\,[3 - \cos 2(N_+ t)] + \epsilon^2\frac{\eta\sqrt{\pi}}{\sigma}\mathrm{e}^{-q^2/4\sigma^2} \\ &\quad\times 2CA\left(\frac{\eta\gamma}{Q_2}\right)^{1/2}\mathrm{sech}\,\Theta_0[\omega_+^{-1}\cos(N_+ + \omega)t \\ &\quad+ \omega_-^{-1}\cos(N_- + \omega)t] + \mathcal{O}(\epsilon^3), \end{aligned} \tag{24}$$

where

$$\Theta(z,t) = \epsilon\frac{v_e\sqrt{1-2g}}{2P_1}\left(qz - (V_g + \epsilon\frac{v_e}{P_1})t\right), \tag{25}$$

$$\Theta_0(t) = \epsilon^2\frac{v_e\sqrt{1-2g}}{2P_1}t, \tag{26}$$

$$Q = q + \epsilon\frac{v_e}{2P_1}, \tag{27}$$

$$M = \omega + \epsilon\frac{v_e}{2P_1}\left[V_g + \epsilon\left(\frac{g v_e}{P_1}\right)\right], \tag{28}$$

$$N_\pm = \pm\sqrt{W} + \epsilon^2 g\frac{v_e^2}{2P_1^2}, \tag{29}$$

$$\Lambda(z) = \frac{\mu\eta\gamma}{V_{(z)}Q_2}\mathrm{e}^{-\sigma^2 z^2}, \tag{30}$$

$$\omega_\pm = \omega^2 \pm 2\omega\sqrt{W}. \tag{31}$$

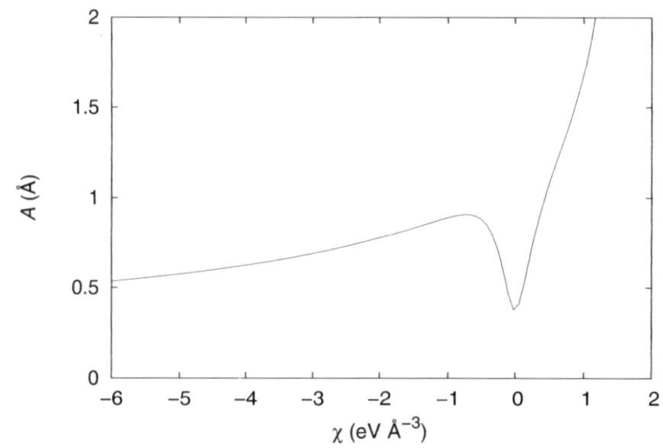

Figure 2. Relation between amplitude A and the coupling constant.

One can see from (23) that if the coupling $\chi = 0$, then $\Lambda(z) = 0$. Hence $y(z,t)$ will be identical to the Peyrard-Bishop breather solution [22], while $x(t) = 0$ as if there is no interacting protein.

We take and adjust the PB parameters from [25],

$$\begin{aligned} \alpha &= 1.2\sqrt{2}\,\text{Å}^{-1}, \quad D = 0.07\,\text{eV}, \quad k = 12\,\text{N/m}, \\ q &= 0.18\,\text{Å}^{-1}, \quad g = 0.47, \quad v_e = 1888\,\text{m/s}. \end{aligned} \tag{32}$$

Here we take constant D for simplicity. The length between two base pairs is $a = 3.4\,\text{Å}$ and the nucleotide mass is $m = 5.1 \times 10^{-25}$ kg. Here the value of q corresponds to a wavelength covering 10 base pairs. We take $E = D$ and $\beta = \alpha$ as the connecting hydrogen bonds between adenine and glutamine in the base pairs; they are the same as those connecting A-T. We interpret the decay factor $\sigma/\sqrt{2}$ as the inverse width of the protein. An α-helix protein is 12 Å in diameter [26], hence $\sigma = 0.117\,\text{Å}^{-1}$. The protein effective mass M is rather free because any proteins could have an arbitrary count of amino acid sequences. Nevertheless, in this case we take it as a glutamic acid weight, $M = 2.47 \times 10^{-24}$ kg. We are left with a free parameter, the coupling constant χ, whose value should have significance in the dynamics of the interaction.

3. Discussion and conclusions

To discuss the model relevance with biological reality, we investigate the restrictions of our introduced coupling constant χ and its ramification with the amplitudes. The value of the coupling is crucial for the complex interaction and should have several interpretations. To see this, let us consider (19) by taking $z = z_0$; we get the restriction

$$Q_1 Q_2 - 3\mu\eta\gamma > 0, \tag{33}$$

or more explicitly, recall that $\mathcal{X} = \chi/a$,

$$Q_1 C + (Q_1 D - A)\mathcal{X}^2 - B\mathcal{X}^3 > 0 \tag{34}$$

where

$$A = \frac{9\sqrt{\pi}}{M\rho\beta^2\sigma}, \quad B = \frac{6\sqrt{\pi}}{M\alpha(\omega^2 - 4W)} e^{-q^2/4\sigma^2}, \tag{35}$$

$$C = 4W, \qquad D = \frac{15\sqrt{\pi/2}}{12M\rho\beta^2\sigma^2 V_{(z)}}. \tag{36}$$

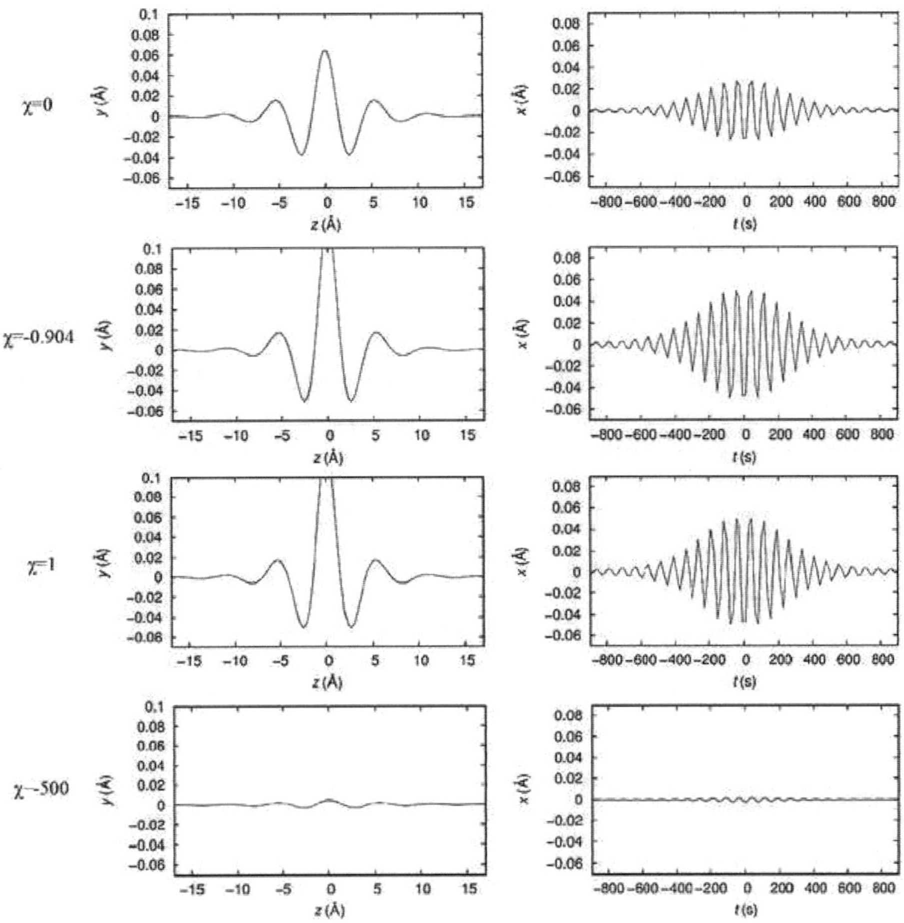

Figure 3. Plot of the solutions with coupling constant $\chi = 0$, $\chi = -0.904$ (local maximum), $\chi = 1$ (approaching singularity) and $\chi = -500$ (highly negative).

If $Q_1D - A > 0$ then we require

$$\sigma < 0.392\,\text{Å}^{-1}, \tag{37}$$

meaning the protein diameter $\sqrt{2}/\sigma > 3.59\,\text{Å}$, which is obviously the case because it is barely a base-pair length. Now we are left with B whose sign depends on $\omega^2 - 4W$. The positive and negative case are

$$M > 8\beta^2 E/\omega^2, \quad M < 8\beta^2 E/\omega^2, \tag{38}$$

respectively. A positive B will produce a relation between the amplitude A and coupling χ as in figure 2; the main property is that $A \to 0$ for highly negative χ. In contrast, a negative B will vertically mirror the relation, i.e., $A \to 0$ for highly positive χ. Here our choice falls in the positive case. We will discuss the four interesting values of χ given in figure 2: zero, local maximum, approaching singularity and highly negative.

For $\chi = 0$ it is shown in (23), figure 3 and figure 4 that the solution is identical to the PB breather when the H-bonds between protein and DNA are not shifted. However, if we take χ near zero then the DNA is open, while the protein side chain is doing a local oscillation. The interaction induces a local base pairs opening to let the protein recognize a specific sequence of base pairs, while previously the bases are being hindered from outside world by DNA strand.

IOP Conf. Series: Journal of Physics: Conf. Series **856** (2017) 012005 doi:10.1088/1742-6596/856/1/012005

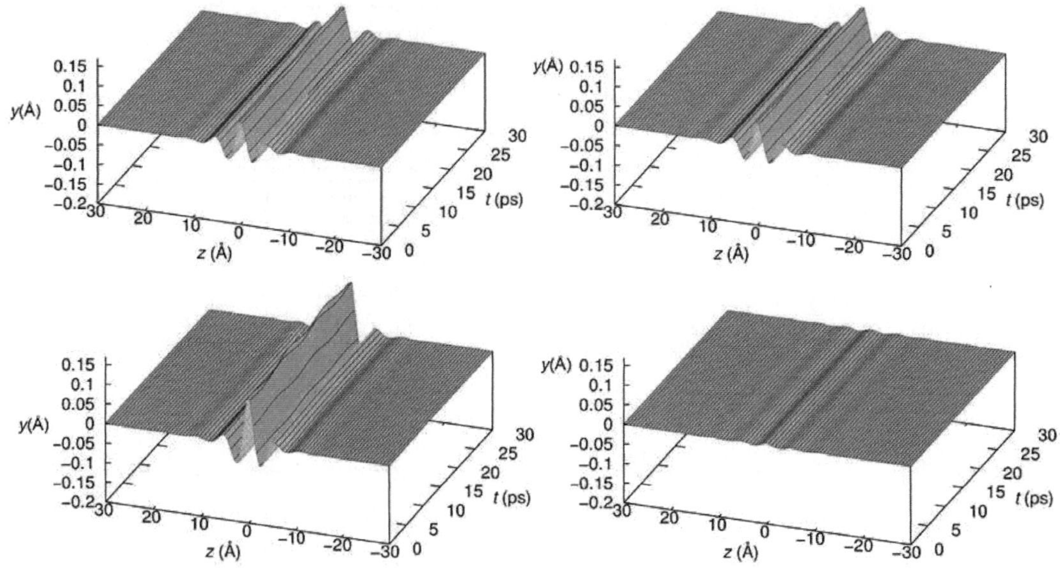

Figure 4. Plot of the strecth y with respect to DNA chain z and time, (top left) $\chi = 0$, (top right) $\chi = -0.904$, (bottom right) $\chi = 1$ and (bottom left) $\chi = -500$.

Our model only contains one-dimensional bond shifts, and thus it has a radial symmetry. It is possible that the bases twist out while being recognized. The propagating breather soliton is interpreted as the mediator of the allosteric transmissions in DNA [2, 4] that facilitate the long-range information transfer between two vastly separated specific DNA-binding proteins.

Our result for small $\chi > 0$ is in agreement with the statistical model in [12] that implies a breather excitation or amplification of an already existing breather. We should restrict χ in the value where the amplitudes vary linearly with χ. In addition, from (23) we find that the wavelength near the protein is reduced by q in the $\Lambda(z)$ term. We predict from (24) that the protein will sustain a small-amplitude oscillation that decays over time, by the slowly varying envelope wave occuring over 1,000 ps. This vibration is mechanically due to the recoil from the base-opening process.

The singular point is due to the square root in (19). We cannot interpret this as a totally denatured or separated DNA strands because the phenomena falls outside our small amplitude approximation. On the other hand, we have not found the significance of the local maximum case of χ other than its relatively high-amplitude property. The local maximum occurs when $Q_1 D - A > 0$. For $Q_1 D \approx A$ or for very small proteins, the local maximum ceases.

The case of highly negative χ is particularly interesting because, contrary to the previous cases, it totally reduces the amplitudes. The protein can repress or close the base pairs within its reach, which is restricted by $f(z)$. This can describe the DNA recombination or the reverse mechanism of denaturation. The naturally occurring base pairs closing is the renaturation process catalyzed by proteins such as RAD1O [27], while a similar mechanism can be engineered by DNA hybridization. The base recombination process is crucial for the polymerase chain reaction which is widely used for DNA testing.

At last, the value of effective mass M is still unclear because it depends on the geometrical features of each type of protein and the binding it conducts. Further investigation is needed because the overall interaction dynamics can be extremely different as the coupling behavior is dependent on M. It is also interesting to study the base pair zippering by protein because it needs to regularize the thermal fluctuation that forces the bases to open again after being closed

IOP Conf. Series: Journal of Physics: Conf. Series **856** (2017) 012005 doi:10.1088/1742-6596/856/1/012005

by the renaturation or hybridization processes.

Acknowledgments

The authors would like to thank Ministry of Research, Technology and Higher Education of Republic of Indonesia for funding Desentralisasi 2016.

References

[1] Xu X, Ge H, Gu C, Gao Y Q, Wang S S, Thio B J R, Hynes J T, Xie X S and Cao J 2013 *J. Phys. Chem.* B **117** 13378
[2] Ptashne M 1986 *Nature* **322** 697
[3] Wang J C and Giaever G N 1988 *Science* **240** 300
[4] Hogan M, Dattagupta N and Crothers D 1979 *Nature* **278** 521
[5] Kim S, Broströmer E, Xing D, Jin J, Chong S, Ge H, Wang S, Gu C, Yang L, Gao Y Q, Su X D, Sun Y and Xie S 2013 *Science* **339** 816
[6] Dwiputra D, Hidayat W and Zen F P 2016 (*Preprint* `1607.01608 [cond-mat.soft]`)
[7] Peyrard M and Bishop A R 1989 *Phys. Rev. Lett.* **62** 2755
[8] Altan-Bonnet G, Libchaber A and Krichevsky O 2003 *Phys. Rev. Lett.* **90** 138101
[9] Sulaiman A, Zen F P, Alatas H and Handoko L T 2012 *Physica* D **241** 1640
[10] Sulaiman A, Zen F P, Alatas H and Handoko L T 2012 *Phys. Scripta* **86** 015802
[11] Hidayat W, Sulaiman A, Viridi S and Zen F 2015 *J. Phys. Chem. Biophys.* **5** 186
[12] Satarić M V and Tuszyński J A 2002 *Phys. Rev.* E **65** 051901
[13] Traverso J J, Manoranjan V S, Bishop A, Rasmussen K Ø and Voulgarakis N K 2015 *Sci. Rep.* **5** 9037
[14] Rohs R, Jin X, West S M, Joshi R, Honig B and Mann R S 2010 *Annu. Rev. Biochem.* **79** 233
[15] Halford S E and Marko J F 2004 *Nucleic Acids Res.* **32** 3040
[16] Von Hippel P H and Berg O G 1986 *P. Natl. Acad. Sci. USA* **83** 1608
[17] Williams M C and Rouzina I 2010 *Biophysics of DNA-Protein Interactions: From Single Molecules to Biological Systems* ed Williams M C and Maher J L (New York: Springer) p 163
[18] Dwiputra D, Hidayat W, Khairani R and Zen F 2016 *J. Phys.: Conf. Ser.* **694** 012076
[19] Dwiputra D, Hidayat W, Khairani R and Zen F 2016 *J. Phys.: Conf. Ser.* **739** 012030
[20] Remoissenet M 1986 *Phys. Rev.* B **33** 2386
[21] Nayfeh A H 2004 *Perturbation Methods* (New York: Wiley)
[22] Peyrard M 2004 *Nonlinearity* **17** R1
[23] Radhakrishnan R and Lakshmanan M 1995 *J. Phys.* A **28** 2683
[24] Hirota R 2004 *The Direct Method in Soliton Theory* (Cambridge: Cambridge University Press)
[25] Zdravković S and Satarić M V 2009 *Phys. Lett.* A **373** 2739
[26] Matsushima N, Danno G i, Takezawa H and Izumi Y 1997 *Biochim. Biophys. Acta* **1339** 14
[27] Sung P, Prakash L and Prakash S 1992 *Nature* **355** 743

IOP Conf. Series: Journal of Physics: Conf. Series **856** (2017) 012006 doi:10.1088/1742-6596/856/1/012006

Power function inflation potential analysis for cosmological model with Gauss-Bonnet term

Getbogi Hikmawan[1], Agus Suroso[1,2] and Freddy P Zen[1,2]

[1] Theoretical Physics Laboratory, THEPI Division, Faculty of Mathematics and Natural Sciences, Institut Teknologi Bandung, Jl. Ganesha 10, Bandung 40132, Indonesia
[2] Indonesia Center of Theoretical and Mathematical Physics (ICTMP), Faculty of Mathematics and Natural Sciences, Institut Teknologi Bandung, Jl. Ganesha 10, Bandung 40132, Indonesia

E-mail: getbogihikmawan@s.itb.ac.id

Abstract. Inflation is still an interesting topic in the study of our universe. An interesting inflation scenario named Gauss-Bonnet inflation proposed without inflation potential has been shown unstable. In this work, we consider the general power function inflation potential, $V(\phi) = m\phi^n$ in the model, then the solution is analyzed according to the inflation scenario. Using the stability condition, inflation potential with m positive and $0 \leq n < 5$ give proper solution for the inflation scenario.

1. Introduction

Inflationary universe scenario at early time by Guth [1] in 1981 has become a possible solution to the horizon and flatness problems in cosmology. There are a lot of ways developed to explain these phenomena, such as slow-roll inflation model [1], Dirac-Born-Infeld (DBI) model [2], ghost inflation model [3], effective field theory [4] and many more. However, those models are still speculative because only few observations show that inflation happens.

Until recently, cosmological models with scalar field have given good solutions for some problems in cosmology, such as for late-time acceleration [5,6], dark matter and dark energy [7,8], and inflation [9, 10]. Kanti *et al* [11, 12] analyzed the cosmological model with nonminimal coupling between scalar field ϕ and Gauss-Bonnet term (GB term), or Einstein-scalar-Gauss-Bonnet theory, without inflation potential and found that the model gives inflationary solution. However, Hikmawan *et al* [13] show that this model is unstable for tensor perturbation modes. In four-dimensional cosmology, GB term arises as a one-loop string correction [14, 15]. Based on this fact, Einstein-scalar-Gauss-Bonnet theory is studied, and it leads to the nonsingular cosmological solutions [16, 17]. The solutions have superinflation phase but generally unstable because of the violation of the weak energy condition [18–22].

The paper is organized as follows. In Section 2, we review the result of Kanti *et al* [11, 12] and Hikmawan *et al* [13] of the Einstein-scalar-Gauss-Bonnet theory and the stability condition required. In Section 3, we include the power function inflation potential, $V(\phi) = m\phi^n$, in the model and numerically solve the field equations for some cases based on the choice of m and n variables. The final section is for conclusion and discussion.

2. Stability condition of cosmological model with Gauss-Bonnet term

In this section, we review the main result of Kanti *et al* [11, 12] and the instability analysis for this model [13]. In particular, we point out the stability condition required for this model to give stable inflation solution.

The action considered in this model is a non-minimally coupling between Gauss-Bonnet term (GB term), R^2_{GB} and a scalar field ϕ via function coupling $f(\phi)$,

$$S = \int d^4x\sqrt{-g}\left[\frac{R}{2\kappa^2} - \frac{1}{2}(\nabla\phi)^2 + \frac{1}{8}f(\phi)R^2_{GB}\right], \tag{1}$$

where $\kappa^2 \equiv 8\pi G$ is the gravitational coupling constant, g is determinant of the metric $g_{\mu\nu}$ and R is the Ricci scalar. GB term is given by

$$R^2_{GB} = R^{\mu\nu\rho\lambda}R_{\mu\nu\lambda\rho} - 4R^{\mu\nu}R_{\mu\nu} + R^2. \tag{2}$$

The variation of this action with respect to the scalar field and the homogeneous and isotropic flat spacetime metric,

$$ds^2 = -dt^2 + a^2(t)\delta_{ij}dx^i dx^j, \tag{3}$$

and assuming further that the scalar field depends solely on the time coordinate, $\phi = \phi(t)$, give explicit field equations for this model,

$$\ddot{\phi} + 3H\dot{\phi} - 3\frac{\partial f}{\partial\phi}H^2(H^2 + \dot{H}) = 0\ , \tag{4}$$

$$6H^2(1 + H\dot{f}) = \dot{\phi}^2\ , \tag{5}$$

$$2(1 + H\dot{f})(H^2 + \dot{H}) + H^2(1 + \ddot{f}) = -\frac{1}{2}\dot{\phi}^2\ , \tag{6}$$

where $H \equiv \dot{a}/a$ is the Hubble parameter and the dot denotes a derivative with respect to time. Kanti *et al* [11, 12] found that the quadratic function $f = \lambda\phi^2$, with λ being a constant, gives rise to the inflationary solutions if $\lambda < 0$ and takes very large value. As the result, the GB term will dominate the Ricci term so that the latter can be neglected.

Because of the existence of ghost, this model is unstable under the tensor perturbation [13]. However, there is a possible way to find a stable cosmological model with the GB term, that is, when the inflation potential $V(\phi)$ is included in the action function:

$$S = \int d^4x\sqrt{-g}\left[\frac{R}{2\kappa^2} - \frac{1}{2}(\nabla\phi)^2 - V(\phi) + \frac{1}{8}f(\phi)R^2_{GB}\right]. \tag{7}$$

Therefore, the field equations can be obtained as

$$\ddot{\phi} + 3H\dot{\phi} - 3f'H^2(H^2 + \dot{H}) + \frac{dV(\phi)}{d\phi} = 0\ , \tag{8}$$

$$6H^2(1 + H\dot{f}) = \dot{\phi}^2 + 2V(\phi)\ , \tag{9}$$

$$2(1 + H\dot{f})(H^2 + \dot{H}) + H^2(1 + \ddot{f}) = -\frac{1}{2}\dot{\phi}^2 + V(\phi)\ . \tag{10}$$

As indicated in [13], the stability condition for cosmological model with GB term is given by

$$\mathcal{Q} \equiv \frac{V(\phi)}{4H^2} - \frac{1}{8}(\frac{\dot{H}}{H^2} + 5)(1 + H\dot{f}) > 0. \tag{11}$$

Therefore, one can find a viable model for cosmological model with GB term by choosing the inflation potential $V(\phi)$.

IOP Conf. Series: Journal of Physics: Conf. Series **856** (2017) 012006 doi:10.1088/1742-6596/856/1/012006

3. Power function inflation potential analysis

Now, let us consider $f(\phi) = \lambda\phi^2$, as in [11, 12], and power function potential

$$V(\phi) = m\phi^n, \tag{12}$$

with n constant integer and m arbitrary constant; thus (8)-(10) can be written as

$$\ddot{\phi} + 3H\dot{\phi} - 6\lambda\phi H^2(H^2 + \dot{H}) + mn\phi^{n-1} = 0 \tag{13}$$

$$6H^2(1 + 2\lambda H\phi\dot{\phi}) = \dot{\phi}^2 + 2m\phi^n\ , \tag{14}$$

$$2(1 + 2\lambda H\phi\dot{\phi})(H^2 + \dot{H}) + H^2(1 + 2\lambda\dot{\phi}^2 + 2\lambda\phi\ddot{\phi}) = -\frac{1}{2}\dot{\phi}^2 + m\phi^n\ . \tag{15}$$

If we combine (14) and (15), we get

$$(5H^2 + 2\dot{H})(1 + 2\lambda H\phi\dot{\phi}) + H^2(1 + 2\lambda\dot{\phi}^2 + 2\lambda\phi\ddot{\phi}) = 2m\phi^n. \tag{16}$$

The stability condition (11) can be written as

$$\mathcal{Q}_m \equiv m\phi^n[12 - 2(\frac{2\dot{H}}{H^2} + 5)] - (\frac{2\dot{H}}{H^2} + 5)\dot{\phi}^2 > 0. \tag{17}$$

We consider two cases from this problem, i.e., $m = -1$ and $m = 1$. After that we do a numerical calculation for each n value until we find the value that makes the model stable. In this numerical calculation, we consider $\lambda = -10^{15}$ to make sure that the GB term is dominant enough.

3.1. $m = -1$ case

If $m = -1$, (13) and (16) can be written as

$$\ddot{\phi} + 3H\dot{\phi} - 6\lambda\phi H^2(H^2 + \dot{H}) - n\phi^{n-1} = 0\ , \tag{18}$$

$$(5H^2 + 2\dot{H})(1 + 2\lambda H\phi\dot{\phi}) + H^2(1 + 2\lambda\dot{\phi}^2 + 2\lambda\phi\ddot{\phi}) = -2\phi^n\ , \tag{19}$$

and stability condition function becomes

$$\mathcal{Q}_{m-} = \phi^n[2(\frac{2\dot{H}}{H^2} + 5) - 12] - (\frac{2\dot{H}}{H^2} + 5)\dot{\phi}^2 > 0. \tag{20}$$

We do numerical calculation for some value of n such as positive, zero and negative, and then plot the solution in e-fold number, $N(t) = \log a(t)$. In doing so, we take $a(0) = 1$, $H(0) = 1$, $\phi(0) = 1$ and $\dot{\phi}(0) =?1$ as boundary conditions because we want to get zero initial e-fold number and the decreasing scalar field after inflation generated. Natural units ($\hbar = c = 1$) are used, so each scale in time coordinate corresponds the time scale in natural units.

We can see in figures 1 and 2 that, for every n value, e-fold number naturally expands in the early times (small t), indicating that inflation happens for this model. For positive n value, the plots of e-fold number for each n value coincide in small t, but they differ in large t where we define large t as $t > 103$ $(\text{GeV})^{-1}$. However, we can see in figures 3 and 4 that for all n values, the model is unstable because $\mathcal{Q}_{m-}$ is always negative, meaning that the stability condition is not satisfied.

IOP Conf. Series: Journal of Physics: Conf. Series **856** (2017) 012006 doi:10.1088/1742-6596/856/1/012006

3.2. $m = 1$ case

If $m = 1$, (13) and (16) can be written as

$$\ddot{\phi} + 3H\dot{\phi} - 6\lambda\phi H^2(H^2 + \dot{H}) + n\phi^{n-1} = 0 \tag{21}$$

$$(5H^2 + 2\dot{H})(1 + 2\lambda H\phi\dot{\phi}) + H^2(1 + 2\lambda\dot{\phi}^2 + 2\lambda\phi\ddot{\phi}) = 2\phi^n \ , \tag{22}$$

and stability condition function becomes

$$\mathcal{Q}_{m+} \equiv \phi^n[12 - 2(\frac{2\dot{H}}{H^2} + 5)] - (\frac{2\dot{H}}{H^2} + 5)\dot{\phi}^2 > 0. \tag{23}$$

We do same analysis as in negative m case.

As shown in figures 5 and 6, the plot of e-fold number as a function of time gives similar result as the negative m case. However, the difference with the negative m case comes from the stability condition. We can see from figure 7 that, for negative n, the model is unstable, but for zero and positive n (figures 8 and 9) there exist some stable cases, that is, for $0 \leq n < 5$. For $n \geq 5$ case, we can see that initially $\mathcal{Q}_{m+}$ is positive, but as t increases, the value becomes negative.

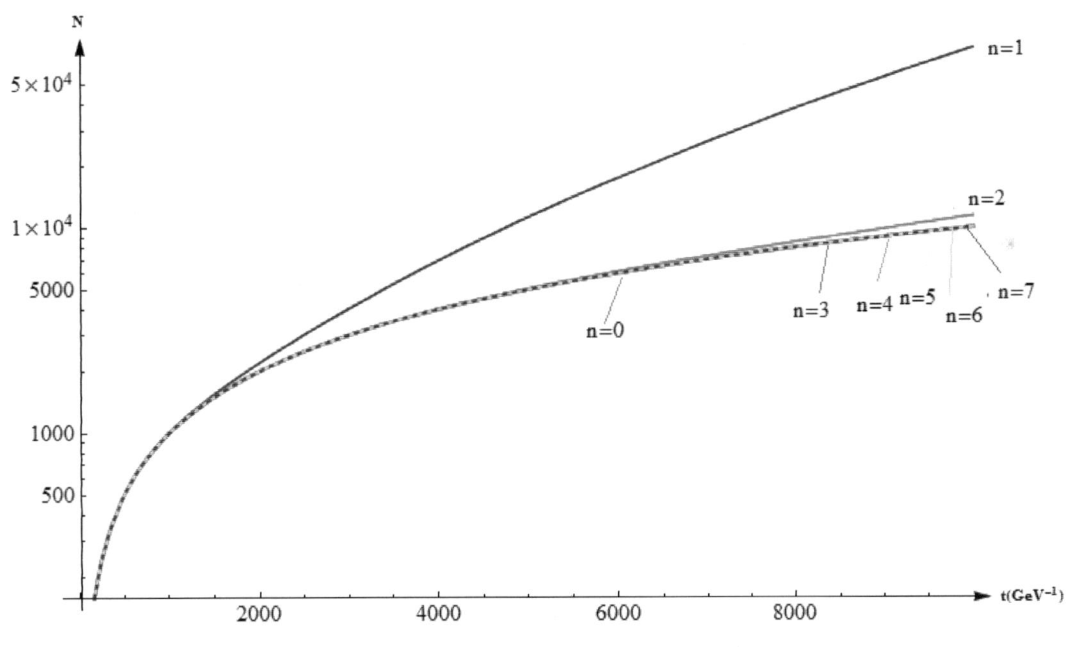

Figure 1. Plot of e-fold number, $N(t) = \log a(t)$ as a function of time for various positive values of n for $m = -1$ case.

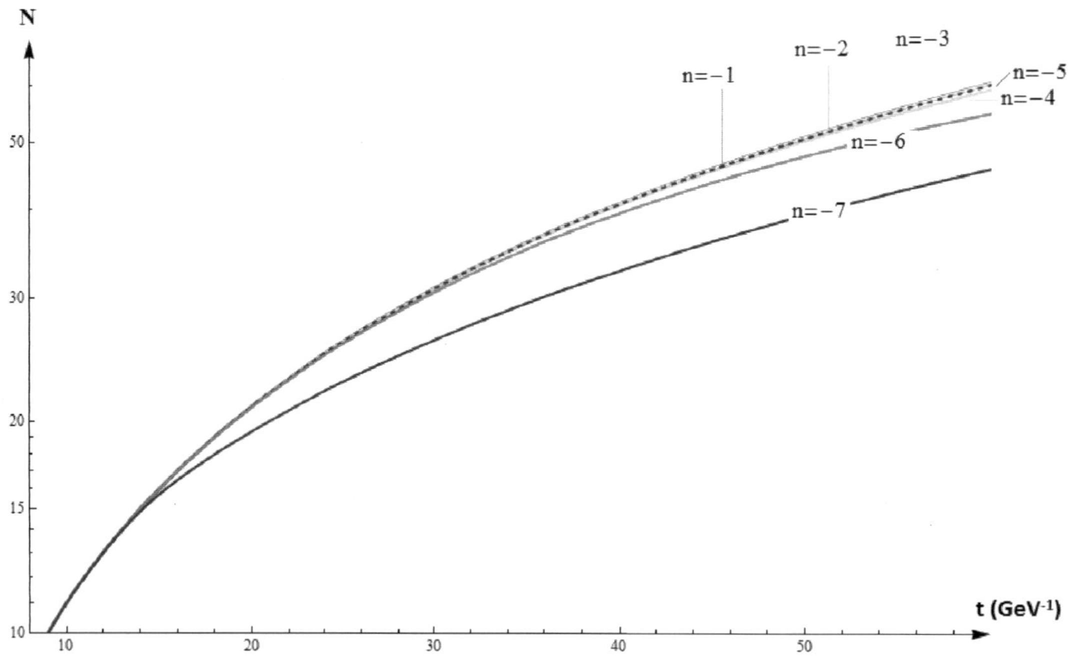

Figure 2. Plot of e-fold number, $N(t) = \log a(t)$ as a function of time for negative values of n for $m = -1$ case.

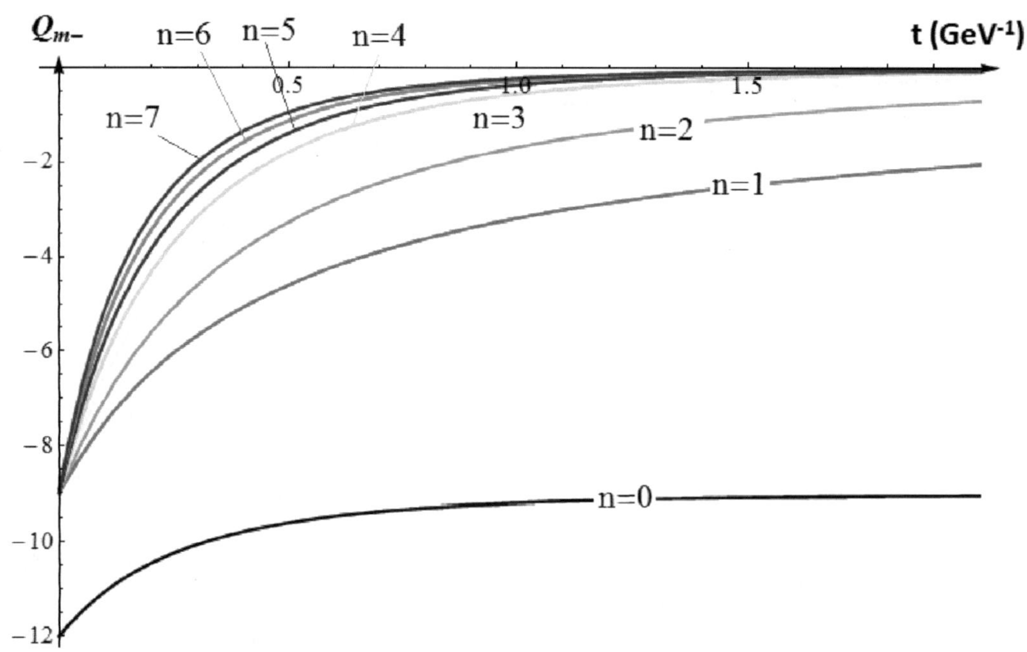

Figure 3. Plot of $\mathcal{Q}_{m^-}$ of the model as a function of time for various positive values of n for $m = -1$ case.

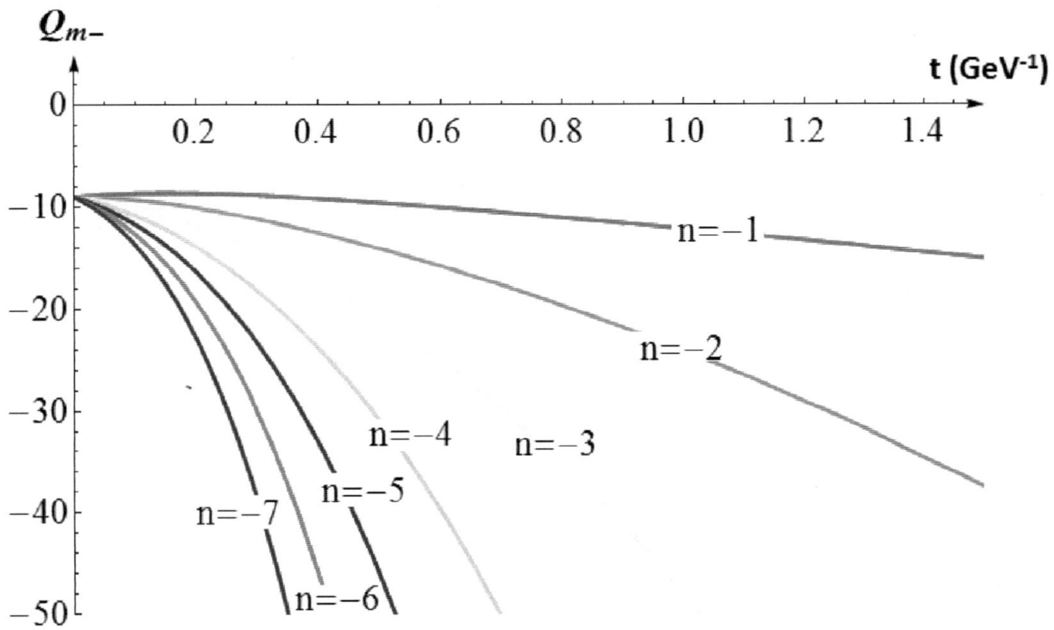

Figure 4. Plot of $\mathcal{Q}_{m^-}$ of the model as a function of time for various negative values of n for $m = -1$ case.

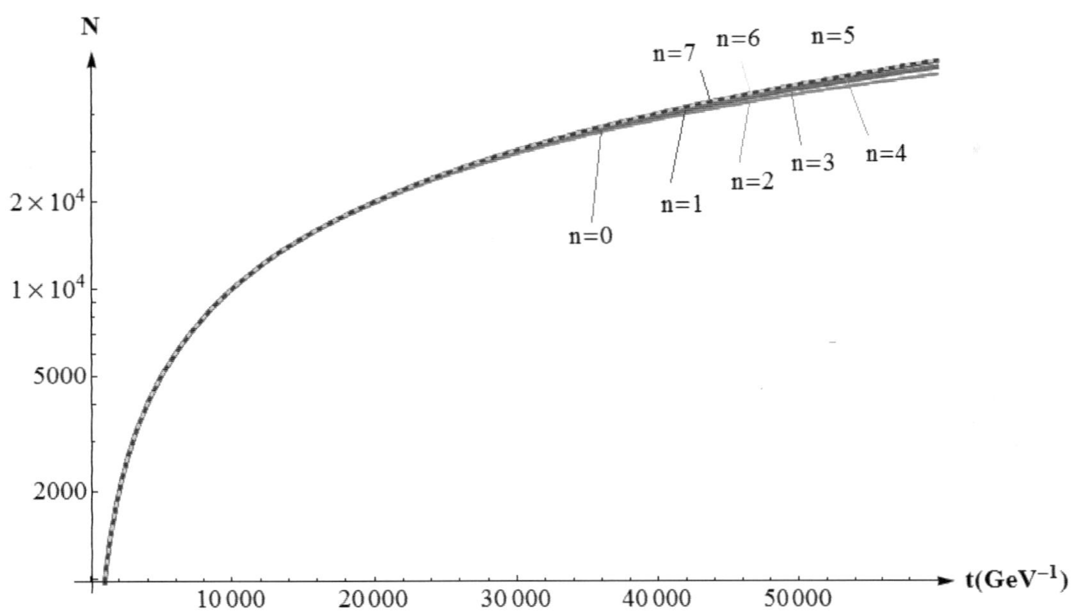

Figure 5. Plot of e-fold number, $N(t) = \log a(t)$ as a function of time for various positive values of n for $m = 1$ case.

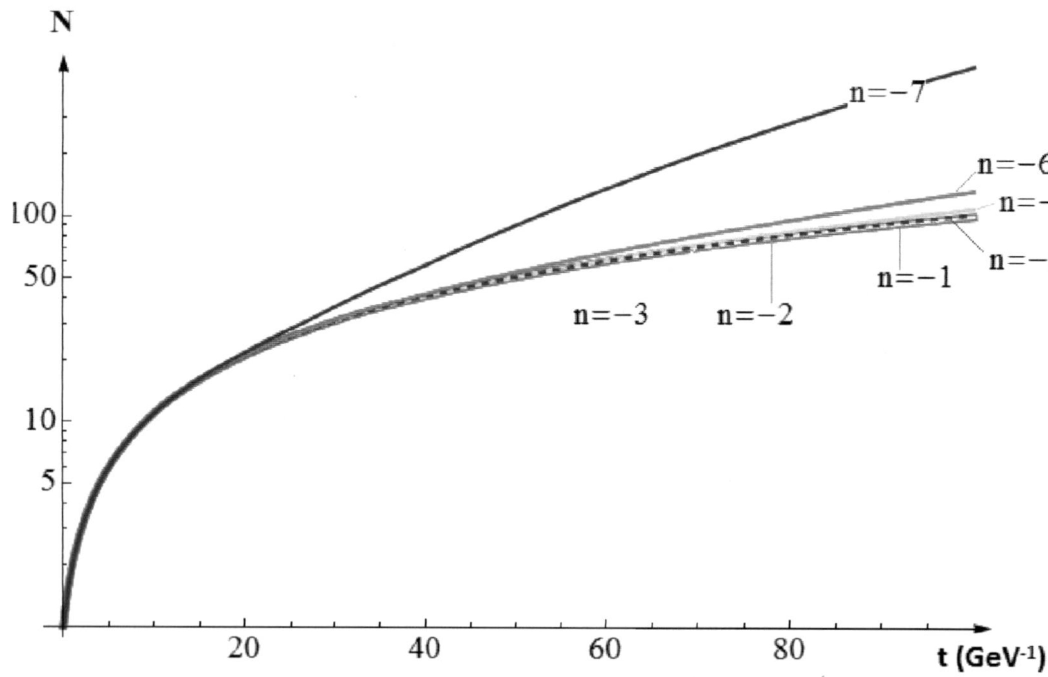

Figure 6. Plot of e-fold number, $N(t) = \log a(t)$ as a function of time for various negative values of n for $m = 1$ case.

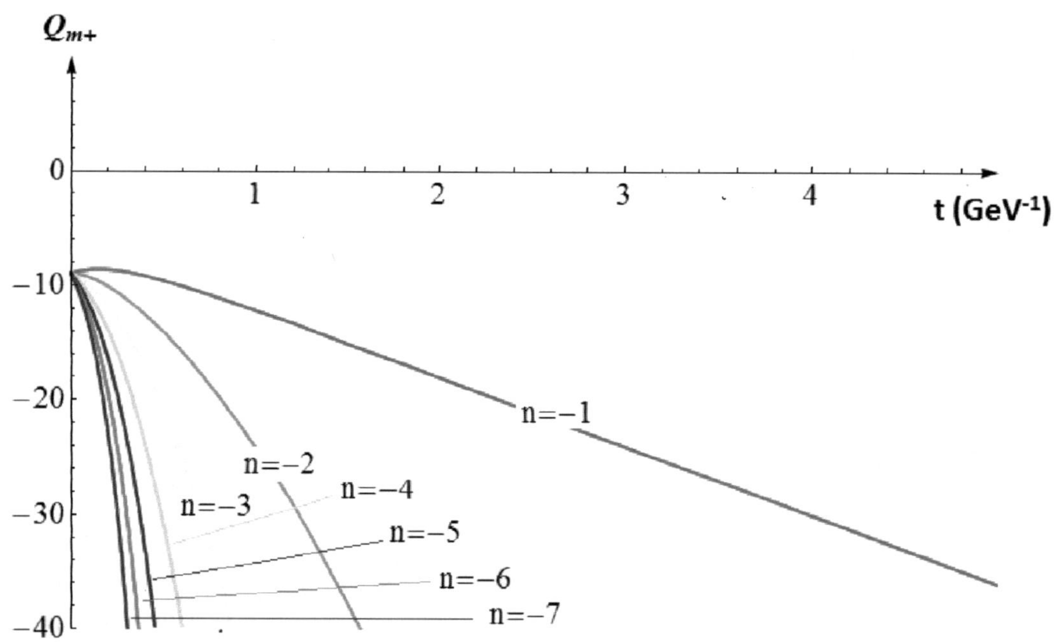

Figure 7. Plot of Q_{m+} of the model as a function of time for various negative values of n for $m = 1$ case

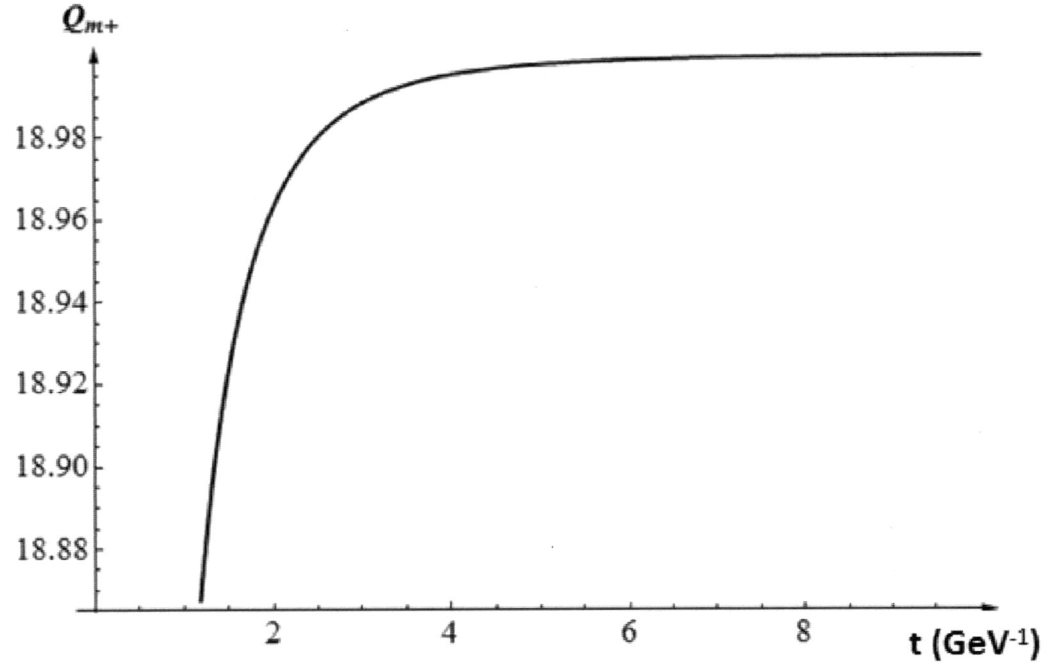

Figure 8. Plot of $\mathcal{Q}_{m^-}$ of the model as a function of time for $n = 0$ for $m = 1$ case

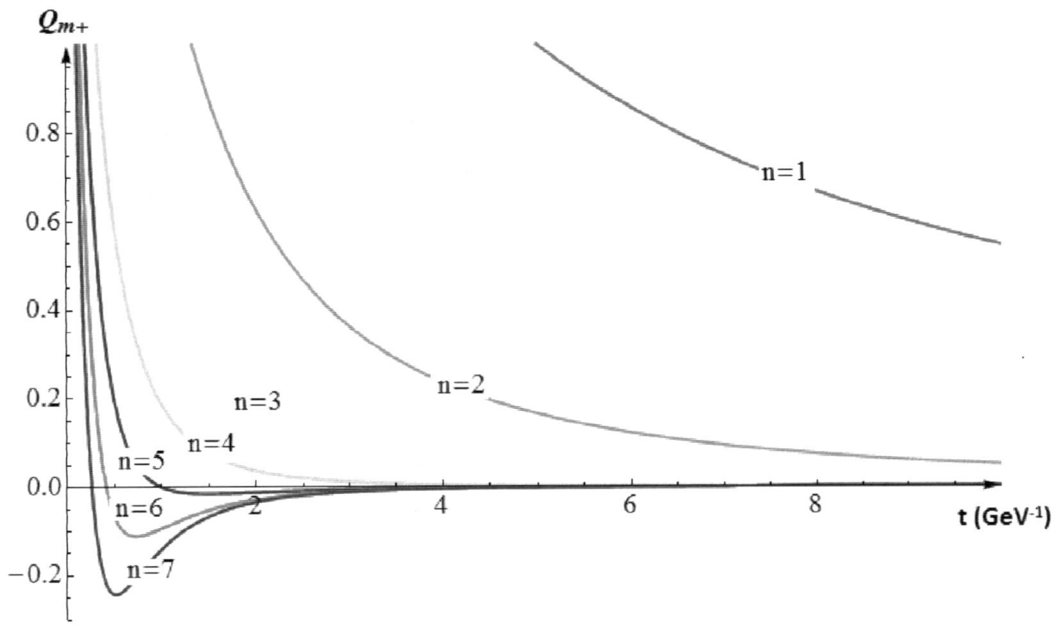

Figure 9. Plot of $\mathcal{Q}_{m+}$ of the model as a function of time for various positive values of n for $m = 1$ case

4. Discussion and conclusion

We have analyzed the influence of power function inflation potential in cosmological model with Gauss-Bonnet term (GB term). In this model, in the case of quadratic coupling between GB term and scalar field, inflation occurs naturally for each case considered. Using the stability condition [13], it can be seen that there exist some stable cases, that is, for $0 \leq n < 5$ with $m = 1$. Although, for both cases, the e-fold number for each n value coincides in small t, but differs in large t. This can happen perhaps because the difference between n values in the inflation potential affects the e-fold number later in large t. However, it does not change the result that the addition of power function inflation potential can be a way to find the stable cosmological model with the GB term. There are other works too that can show there are stable inflationary solutions [23–27], and the result of this works can make intriguing motivation to study Einstein-scalar-Gauss-Bonnet theory in the cosmological context.

Acknowledgments

This work was supported by "Riset Inovasi KK ITB 2016", "Riset Desentralisasi ITB 2016" and "Riset PMDSU 2016" from Ministry of Research, Technology and Higher Education of the Republic of Indonesia.

References

[1] Guth A H 1981 *Phys. Rev.* D **23** 347
[2] Alishahiha M, Silverstein E and Tong D 2004 *Phys. Rev.* D **70** 123505
[3] Arkani-Hamed N, Creminelli P, Mukohyama S and Zaldarriaga M 2004 *J. Cosmol. Astropart. Phys.* JCAP04(2004)001
[4] Cheung C, Creminelli P, Fitzpatrick A L, Kaplan J and Senatore L 2008 *J. High Energy Phys.* JHEP03(2008)014
[5] Caldwell R R, Dave R and Steinhardt P J 1998 *Phys. Rev. Lett.* **80** 1582
[6] Sushkov S V 2009 *Phys. Rev.* D **80** 103505
[7] Zen F P, Arianto, Gunara B E, Triyanta and Purwanto A 2009 *Eur. Phys. J.* C **63** 477
[8] Amendola L 1993 *Phys. Lett.* B **301** 175
[9] Capozziello S, Lambiase G and Schmidt H J 2000 *Annalen. Phys.* **9** 39
[10] Tsujikawa S 2012 *Phys. Rev.* D **85** 083518
[11] Kanti P, Gannouji R and Dadhich N 2015 *Phys. Rev.* D **92** 041302
[12] Kanti P, Gannouji R and Dadhich N 2015 *Phys. Rev.* D **92** 083524
[13] Hikmawan G, Soda J, Suroso A and Zen F P 2016 *Phys. Rev.* D **93** 068301
[14] Antoniadis I, Gava E and Narain K S 1992 *Phys. Lett.* B **283** 209
[15] Antoniadis I, Gava E and Narain K S 1992 *Nucl. Phys.* B **383** 93
[16] Easther R and Maeda K i 1996 *Phys. Rev.* D **54** 7252
[17] Kawai S and Soda J 1999 *Phys. Rev.* D **59** 063506
[18] Kawai S, Sakagami M and Soda J 1997 Perturbative analysis of nonsingular cosmological model (*Preprint* gr-qc/9901065)
[19] Kawai S, Sakagami M and Soda J 1998 *Phys. Lett.* B **437** 284
[20] Soda J, Sakagami M and Kawai S 1998 Novel instability in superstring cosmology (*Preprint* gr-qc/9807056)
[21] Kawai S and Soda J 1999 *Phys. Lett.* B **460** 41
[22] Kawai S and Soda J 1999 Structure formation from nonsingular kinetic inflation (*Preprint* gr-qc/9906046)
[23] Neupane I P 2007 Constraints on Gauss-Bonnet cosmologies (*Preprint* arXiv:0711.3234 [hep-th])
[24] Satoh M, Kanno S and Soda J 2008 *Phys. Rev.* D **77** 023526
[25] Satoh M and Soda J 2008 *J. Cosmol. Astropart. Phys.* JCAP09(2008)019
[26] Guo Z K and Schwarz D J 2009 *Phys. Rev.* D **80** 063523
[27] Satoh M 2010 *J. Cosmol. Astropart. Phys.* JCAP11(2010)024

IOP Conf. Series: Journal of Physics: Conf. Series **856** (2017) 012007 doi:10.1088/1742-6596/856/1/012007

Jin–Xin relaxation method used to solve the one-dimensional inviscid Burgers equation

Sudi Mungkasi[1], Bernadeta Wuri Harini[2] and Leo Hari Wiryanto[3]

[1] Department of Mathematics, Faculty of Science and Technology, Sanata Dharma University, Mrican, Tromol Pos 29, Yogyakarta 55002, Indonesia

[2] Department of Electrical Engineering, Faculty of Science and Technology, Sanata Dharma University, Mrican, Tromol Pos 29, Yogyakarta 55002, Indonesia

[3] Department of Mathematics, Faculty of Mathematics and Natural Sciences, Bandung Institute of Technology, Jalan Ganesha 10, Bandung 40132, Indonesia

E-mail: `sudi@usd.ac.id`, `wuribernard@usd.ac.id`, `leo@math.itb.ac.id`

Abstract. We consider the Burgers equation and seek for its numerical solutions. The relaxation method of Jin and Xin, called the Jin–Xin relaxation method, is tested to solve the Burgers equation. We find that the Jin–Xin relaxation method solves the Burgers equation successfully. In addition, the Jin–Xin relaxation method holds the analytical properties of the Burgers equation better than the standard Lax–Friedrichs finite-volume method does.

1. Introduction

Conservation laws govern a number of physical and mathematical problems. In physics, they relate to transport of conserved quantities, such as mass, momentum and energy. In mathematics, they may occur as a constraint in optimal control problems and other optimization problems, for example, minimizing costs, maximizing revenue, etc.

Some previous works related to conservation laws are available in the literature. Conservation laws were successfully applied to simulate real world problems, such as floods and tsunamis as presented by Mungkasi and Darmawan [1]. Investigation on Adjoint IMEX-based schemes for control problems was conducted by Banda and Herty [2] using the relaxation method of Jin and Xin [3] (the Jin–Xin relaxation method). Yohana [4], as well as Yohana and Banda [5], extends the investigation to the Euler equations of gas dynamics employing the work of Armijo [6]. Even though the Jin–Xin relaxation method has been implemented successfully by Jin and Xin [3] themselves, as well as by the authors of [2,4,5], to solve some problems, to our knowledge, there is no publication testing the method to solve the one-dimensional inviscid Burgers equation.

In this paper we solve the one-dimensional inviscid Burgers equation using the Jin–Xin relaxation method. Note that the Burgers equation is a member of the class of conservation laws. Our work complements the aforementioned literature. We also solve the Burgers equation using the standard Lax-Friedrichs finite-volume method, as a benchmark, to compare the results with.

This paper is organized as follows. We write the problem formulation of the Jin–Xin relaxation system in Section 2. The numerical scheme to solve the problem is given in Section 3. Numerical results and discussion are presented in Section 4. Finally, we draw some concluding remarks in Section 5.

IOP Conf. Series: Journal of Physics: Conf. Series **856** (2017) 012007 doi:10.1088/1742-6596/856/1/012007

2. Problem formulation

In this section, we describe the problem that we want to deal with. We write the mathematical model under consideration first and the approach to solve the model afterward. We assume that the model is one-dimensional and scalar.

First, let us recall the general mathematical model that we want to solve. The general mathematical model governing the one-dimensional scalar conservation law is

$$u_t + f(u)_x = 0, \tag{1}$$

for the time variable $t \in [0, \infty)$ and the space variable $x \in (-\infty, \infty)$ with the initial condition

$$u\,(x, t = 0) = u_0(x). \tag{2}$$

Here $u = u(x,t)$ denotes the conserved quantity and $f\,(u)$ represents the flux function. The function u_0 gives the initial condition of the conserved quantity. The conserved quantity can be mass, momentum or energy. Equation (1) is generally nonlinear. Note that, taking $f(u) = \frac{1}{2}u^2$, (1) becomes the one-dimensional inviscid Burgers equation [7].

Now, to solve (1) with the initial condition (2) let us recall the relaxation approach of Jin and Xin [3], that is, the Jin–Xin relaxation method. Jin and Xin [3] proposed that (1) and (2) be approximated by a relaxation system, which is a linear system of partial differential equations. The Jin–Xin relaxation system for the mathematical model (1) is [3]

$$u_t + v_x = 0, \tag{3}$$

$$v_t + au_x = -\frac{1}{\epsilon}\left(v - f(u)\right), \tag{4}$$

where the initial conditions relating to (2) are

$$u(x, 0) = u_0(x), \tag{5}$$

$$v(x, 0) = f\left(u_0(x)\right). \tag{6}$$

Here $v = v(x,t)$ is a dummy variable and a is a positive constant satisfying

$$-\sqrt{a} \leq f'(u) \leq \sqrt{a}, \tag{7}$$

for all u. The parameter ϵ is positive and called the relaxation rate. When the solution is smooth, we eliminate variable v from (3) and (4) to obtain

$$u_t + f(u)_x = \epsilon\left(au_{xx} - u_{tt}\right). \tag{8}$$

For $\epsilon \to 0^+$, we observe that (8) tends to (1). That is, for $\epsilon \to 0^+$, (3) and (4) simultaneously tend to (1). Here we have approximated the nonlinear equation (1) using the system of linear equations (3) and (4) for $\epsilon \to 0^+$. However, if we take $\epsilon = 0$, then (3) and (4) are not defined. Ideally ϵ is a small positive number, which is close to zero. More details about the Jin–Xin relaxation system can be found in the work of Jin and Xin [3].

3. Numerical method

We write the numerical scheme to solve (3) and (4), that is, the relaxation system of (1).

Let us assume that the space and time domains are discretized in the finite volume framework. The space domain is discretized uniformly with the cell width h. (Readers interested in nonuniform space discretization can consult the work of Mungkasi [8].) The time domain is discretized uniformly with the time step k. The notation u_j^n is for pointing the approximate value of u in the j-th cell at the n-th time step.

In this paper we use the upwind numerical technique as follows [3]. Discretizing (3) and (4) with respect to space and time domains, we obtain the one-step conservative explicit scheme:

$$\frac{u_j^{n+1} - u_j^n}{k} + \frac{v_{j+1/2}^n - v_{j-1/2}^n}{h} = 0, \tag{9}$$

$$\frac{v_j^{n+1} - v_j^n}{k} + a\frac{u_{j+1/2}^n - u_{j-1/2}^n}{h} = -\frac{1}{\epsilon}\left(v_j^n - f\left(u_j^n\right)\right). \tag{10}$$

Once again, a is a positive constant satisfying (7), so we can take $a = f'(u)^2$. Jin and Xin [3] derived the first-order upwind relaxing fluxes, which are given by

$$u_{j+1/2}^n = \frac{1}{2}\left(u_j^n + u_{j+1}^n\right) - \frac{1}{2\sqrt{a}}\left(v_{j+1}^n - v_j^n\right), \tag{11}$$

$$v_{j+1/2}^n = \frac{1}{2}\left(v_j^n + v_{j+1}^n\right) - \frac{\sqrt{a}}{2}\left(u_{j+1}^n - u_j^n\right). \tag{12}$$

Using the upwind fluxes (11) and (12) in the one-step conservative explicit schemes (9) and (10) we obtain the upwind numerical method for (3) and (4) as

$$\frac{u_j^{n+1} - u_j^n}{k} + \frac{1}{2h}\left(v_{j+1}^n - v_{j-1}^n\right) - \frac{\sqrt{a}}{2h}\left(u_{j+1}^n - 2u_j^n + u_{j-1}^n\right) = 0, \tag{13}$$

$$\frac{v_j^{n+1} - v_j^n}{k} + \frac{a}{2h}\left(u_{j+1}^n - u_{j-1}^n\right) - \frac{\sqrt{a}}{2h}\left(v_{j+1}^n - 2v_j^n + v_{j-1}^n\right) = -\frac{1}{\epsilon}\left(v_j^n - f\left(u_j^n\right)\right). \tag{14}$$

Therefore, given that we know the quantities u_j^n and v_j^n at the n-th time step for all j cells, using (13) and (14) we can compute the quantities and at the $(n+1)$-th time step for all j cells.

This can be done by iterations. Starting from the initial condition (2), we have (5), so we obtain u_j^0. We also have (6), so we obtain v_j^0. Then the values of u_j^1, v_j^1, u_j^2, v_j^2, u_j^3, v_j^3, ... are also obtained for all j using (13) and (14).

4. Numerical results

In this section we present results of our research. The Jin–Xin relaxation method is applied to solve the inviscid Burgers equation. The results of this method are analyzed if they hold some analytical properties of the Burgers equation. We also compare the results of the Jin–Xin relaxation method to those of the standard Lax-Friedrichs finite-volume method to show that the Jin–Xin relaxation method performs better. All quantities are assumed to be in the SI units, so we do not need to write their units.

As we have mentioned in Section 2, let us take $f(u) = \frac{1}{2}u^2$. Equation (1) becomes the nonlinear inviscid Burgers equation

$$u_t + \left(\frac{1}{2}u^2\right)_x = 0. \tag{15}$$

IOP Conf. Series: Journal of Physics: Conf. Series **856** (2017) 012007 doi:10.1088/1742-6596/856/1/012007

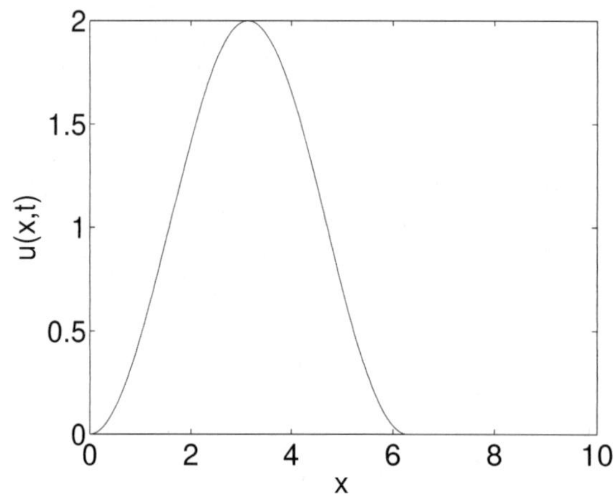

Figure 1. Initial condition for the Burgers equation at time $t = 0$.

For simulations, we set up the followings. We consider the space domain $0 \leq x \leq 10$. The initial condition is given by

$$u_0(x) = \begin{cases} 1 - \cos x, & \text{if} \quad 0 < x \leq 2\pi, \\ \\ 0, & \text{if} \quad 2\pi < x \leq 10. \end{cases} \tag{16}$$

The boundary condition is

$$u(0, t) = u(10, t) = 0 \tag{17}$$

for all $t > 0$. For all simulations, we take the cell width to be $h = 10^{-2}$ and the time step to be $k = 10^{-3}$. This time-step value is chosen so, in order that methods are stable. The Jin–Xin relaxation rate is taken as $\epsilon = 10^{-3}$.

Let us collect some analytical properties of our simulation setting. With the initial condition given by (16), the maximum height of the wave is 2, as shown in figure 1. Due to the characteristics of the Burgers equation, the wave propagates from the left to the right direction. Based on the formulation of LeVeque [9], the wave will form a shock discontinuity starting at time $t_s = -1/\min\{u_0'(x)\}$, that is, $t_s = 1$. Therefore, a shock wave will propagate from the left to the right at time $t > 1$. When the shock occurs, the wave loses some energy, so the wave height gets smaller over time. Using these properties, we shall analyze our numerical results.

Representations of Jin–Xin relaxation results are shown in figure 2. From our simulation, we observe that the wave height is kept at the value of 2 until time $t_s = 1$, as given in figure 2(a). Then for time $t > 1$, a shock occurs and the wave height gets smaller than 2. We also notice that the shock is sharply resolved by the Jin–Xin relaxation method, as illustrated in figure 2(b).

In this section we also record numerical results of the standard Lax-Friedrichs finite-volume method applied directly to the Burgers equation (15). Readers interested in this finite-volume method are referred to the work of LeVeque [10]. In the Lax-Friedrichs finite volume simulation, we use the same initial and boundary conditions as well as the same values of the cell width and time step, as mentioned before. Representations of Lax-Friedrichs finite volume results are shown in figure 3. The Lax-Friedrichs finite-volume method cannot preserve the wave height at the value of 2 at time $t_s = 1$, as given in figure 3(a). In addition, the Lax-Friedrichs finite-volume method cannot sharply resolve the shock wave for time $t > 1$, as shown in figure 3(b).

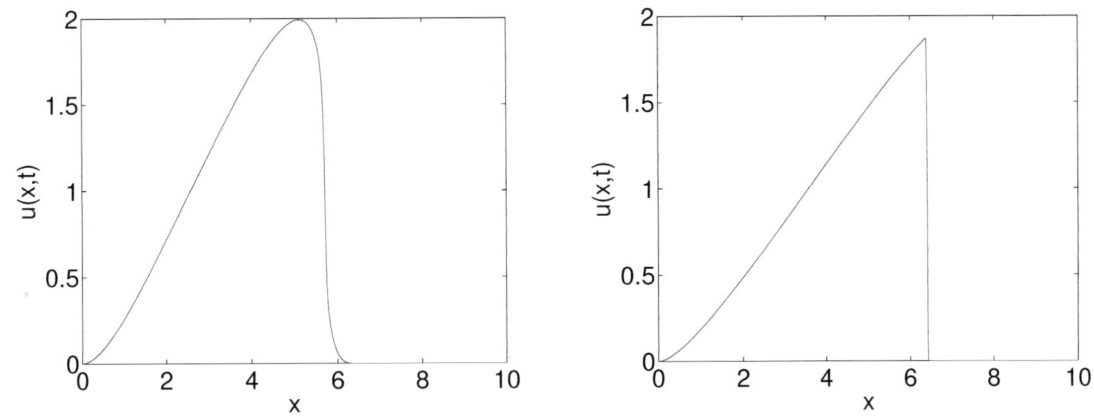

(a) The Jin–Xin solution at time $t = 1$. (b) The Jin–Xin solution at time $t = 2$.

Figure 2. Solutions to the Burgers equation using the Jin–Xin relaxation method.

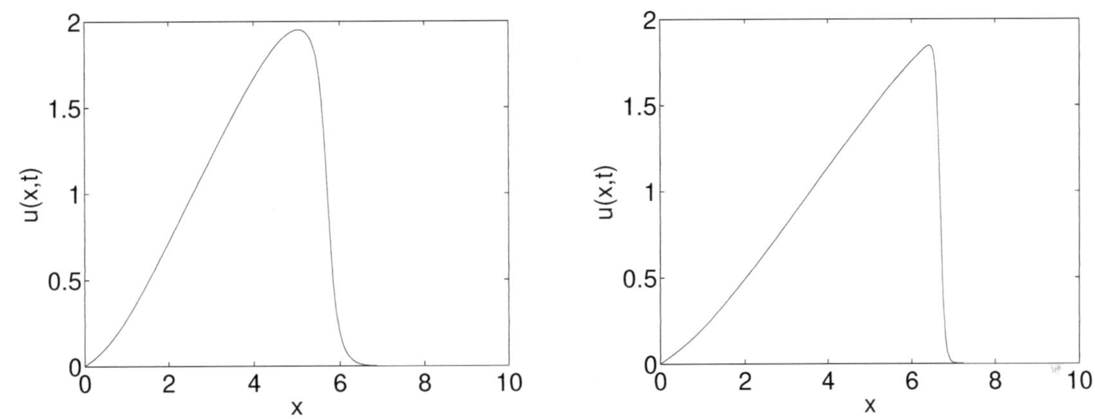

(a) The Lax-Friedrichs solution at time $t = 1$. (b) The Lax-Friedrichs solution at time $t = 2$.

Figure 3. Solutions to the Burgers equation using the Lax-Friedrichs finite-volume method.

We can now compare the results of the Jin–Xin relaxation method and those of the Lax-Friedrichs finite-volume method. Comparing figure 2(a) and figure 3(a), we obtain that the Jin–Xin relaxation method holds the analytical property of the wave height better than the Lax-Friedrichs finite-volume method does. Furthermore, comparing figure 2(b) and figure 3(b), we notice that the Jin–Xin relaxation method holds the analytical property of the shock discontinuity better than the Lax-Friedrichs finite-volume method does.

5. Conclusion

We have solved the inviscid Burgers equation using the Jin–Xin relaxation method successfully. Our results in this paper complement the test cases already available in the literature regarding the method. The method is able to resolve smooth and nonsmooth solutions. In addition, the Jin–Xin relaxation method has better performance than the standard Lax-Friedrichs finite-volume method does.

Acknowledgments

This work was financially supported by Sanata Dharma University and a research grant from *Direktorat Riset dan Pengabdian Masyarakat* (DRPM) of Ministry of Research, Technology and Higher Education of the Republic of Indonesia year 2016.

References

[1] Mungkasi S and Darmawan J B B 2015 *Proc. 4th Int. Conf. on Soft Computing, Intelligent System and Information Technology (Bali)* vol 516 ed Intan R *et al* (Springer-Verlag) p 469
[2] Banda M K and Herty M 2012 *Comput. Optim. Appl.* **51** 909
[3] Jin S and Xin Z 1995 *Commun. Pure Appl. Math.* **48** 235
[4] Yohana E 2012 MSc Thesis University of the Witwatersrand at Johannesburg
[5] Yohana E M and Banda M K 2016 *J. Numer. Math.* **24** 45
[6] Armijo L 1966 *Pac. J. Math.* **16** 1
[7] Mungkasi S 2015 *IOP Conf. Ser.: Mater. Sci. Eng.* **78** 012031
[8] Mungkasi S 2016 *Adv. Math. Phys.* **2016** 7528625
[9] LeVeque R J 1992 *Numerical Methods for Conservation Laws* (Basel: Springer)
[10] LeVeque R J 2002 *Finite Volume Methods for Hyperbolic Problems* (Cambridge: Cambridge University Press)

Compactification stability by global k-monopole in higher dimensional σ-model

Ilham Prasetyo and Handhika S Ramadhan

Departemen Fisika, FMIPA, Universitas Indonesia, Depok 16424, Indonesia

E-mail: ilham.prasetyo51@ui.ac.id, hramad@ui.ac.id

Abstract. We investigate compactification by k-monopole in nonlinear sigma model with kinetic terms $K(X) = -X - \beta^{-2}X^2$. We use both Einstein equations and effective potential from dimensional reduction to determine radius of compactification. The compactification channels can be found by introducing effective potential, and its derivatives shall determine the stability.

1. Introduction

Monopole is a noncontractible point defect due to its second fundamental group being nontrivial, $\pi_2(S^2) = \mathbb{Z}$ [1, 2]. Specifically due to the absence of gauge field, global monopole has divergent energy. It causes a peculiar feature when coupled to gravity: it gives no gravitational force on the surrounding matter, except near the core due to its tiny mass [3]. However, the space around global monopole suffers from deficit solid angle $\Delta \equiv 8\pi G\eta^2$, which makes it non-Euclidean. This deficit angle has a critical value $\eta < \sqrt{1/8\pi G}$. In [4], the authors suggest that when η is a little bigger than this upper bound, two angular dimensions of the space around the global monopole might be compactified into a cylinder. In [5, 6] global monopole was investigated in higher-dimensional spacetime when the staticity assumption is relaxed, i.e., the geometry can then be thought of as a "cigar". In other words, compactification of this type can be seen as nonstatic solution. Numerically, as discussed in [7], regular solutions in four-dimensional space could exist up to $\eta \lesssim \sqrt{3/8\pi G}$. Above that value, the singularity is interpreted in [8, 9] as the appearance of topological inflation.

Nonlinear terms in Lagrangian kinetic term, whose defects can arise from without providing a symmetry-breaking potential term [10], are named in [11] as k-defect. In the case of topological defect from monopoles, they are called in [12] as k-monopole.

In [13, 11] global k-monopole has been studied, and investigations of its gravitational field were done in [12, 14]. Qualitatively, they show that the gravitational property of Barriola-Vilenkin monopole still holds. The difference is their mass can be negative or positive, depending on the specific model of k-term considered. The region outside the monopole can be studied using the vacuum approximation where the Higgs field is approximation, where Higgs field is approximated as $|\phi| \simeq \eta$ so that the analytical solutions can be found. The authors of [15] show some gravitational solutions at the exterior of two different types of global k-monopoles [12].

Here we report our finding for a different model: a sigma model based on [16] with a cosmological constant and nonlinear term based on [11]. We investigate static compactification solutions and its stability by introducing an effective potential.

IOP Conf. Series: Journal of Physics: Conf. Series **856** (2017) 012008 doi:10.1088/1742-6596/856/1/012008

2. Global k-monopole in sigma model

Here we use $(+--...)$ metric g_{MN} $(M,N=0,...,p,\theta^1,...,\theta^{D-2})$ for static $(p+D)$-dimensional spacetime

$$ds^2 = g_{MN}dx^M dx^N = A^2(r)\eta_{\mu\nu}dx^\mu dx^\nu - B^2(r)dr^2 - C^2 d\Omega^2_{D-2}, \tag{1}$$

with $(p+1)$-dimensional flat space and $(D-2)$-sphere denoted by $g_{\mu\nu}dx^\mu dx^\nu = A(r)^2\eta_{\mu\nu}dx^\mu dx^\nu$ $(\mu,\nu=0,...,p)$ and $g_{ij}dx^i dx^j = -C^2 d\Omega^2_{D-2}$ $(i,j=\theta^1,...,\theta^{D-2})$ respectively, C is a real-valued constant and $D\geq 4$. The metric is induced in the action

$$\mathcal{S} = \int d^{p+D}x\,\sqrt{|g|}\left[\frac{R-2\Lambda}{2\kappa} + K(\mathfrak{X}) - \frac{\lambda}{4}\left(\vec{\Phi}^2-\eta^2\right)^2\right], \tag{2}$$

with $K(\mathfrak{X}) = -\mathfrak{X} - \beta^{-2}\mathfrak{X}^2$ and $\mathfrak{X} \equiv -(1/2)g^{MN}\partial_M\vec{\Phi}\cdot\partial_N\vec{\Phi}$. κ is Newton constant in $(p+D)$-dimensional spacetime. The nonlinear term containing β in the kinetic part goes linear when $\beta\to\infty$.

The action, being defined to be invariant under SO(n) transformation, has n degrees of freedom. Our object of interest is nonlinear sigma model with a constraint $\vec{\Phi}^2=\eta^2$, which means that the last term in (2) is just a term multiplied by Lagrange multiplier λ. The scalar field lives in n-dimensional manifold, but now it is constrained to a surface of $(n-1)$-sphere. By this we can introduce its internal basis coordinates in the S^{n-1} by $\vec{\Phi}\equiv\vec{\Phi}(\phi^i)$ with $i=1,...,n-1$. By these constructions (following [16]) we can write a nonlinear σ-model action

$$\mathcal{S} = \int d^{p+D}x\,\sqrt{|g|}\left[\frac{R-2\Lambda}{2\kappa} - X - \beta^{-2}X^2\right], \tag{3}$$

with $X=-(1/2)\eta^2 h_{ij}g^{MN}\partial_M\phi^i\partial_N\phi^j$. Its inner space metric $h_{ij}(\phi^k) = \frac{\partial\vec{\Phi}}{\partial\phi^i}\cdot\frac{\partial\vec{\Phi}}{\partial\phi^j}$ is defined to be dependent on $\phi^1,...,\phi^{n-1}$.

The nonlinear σ-model action then gives us the scalar-field's equation of motion and the energy-momentum tensor for the Einstein equation by the usual variational method

$$\frac{1}{\sqrt{|g|}}\partial_M\left(\sqrt{|g|}(1+2\beta^{-2}X)\eta^2 h_{ij}\partial^M\phi^j\right) = (1+2\beta^{-2}X)\frac{\eta^2}{2}\partial_M\phi^m\partial^M\phi^n\frac{\partial h_{mn}}{\partial\phi^i}, \tag{4}$$

$$T^M_N = \delta^M_N\left[\frac{\Lambda}{\kappa} + X + \beta^{-2}X^2\right] + (1+2\beta^{-2}X)\eta^2 h_{ij}\partial^M\phi^i\partial_N\phi^j. \tag{5}$$

Now it is crucial to choose an appropriate ansatz for the scalar field that satisfies the field equation of motion. This we follow the methods used in [16, 17]. The field's equation of motion is satisfied by choosing the scalar field to be dependent on the angular coordinates $\theta^1,...,\theta^{D-2}$ (the scalar field is now constrained in S^{D-2} with $D\geq 4$, following [6])

$$\phi^i(\theta^i) = \theta^i \tag{6}$$

if and only if we choose its inner metric $h_{ij}(\phi^k)$ to be related to $g_{ij}(C,\theta^k)$ by

$$h_{ij}(\phi^k) = -C^{-2}g_{ij}(C,\theta^k), \tag{7}$$

with C a real-valued constant which gives $X=(D-2)\eta^2/2C^2$. This then gives us the components of energy-momentum tensor

$$T^0_0 = T^r_r = \left[\frac{\Lambda}{\kappa} + X + \beta^{-2}X^2\right], \tag{8}$$

$$T_\theta^\theta = T_0^0 - (1 + 2\beta^{-2}X)\frac{\eta^2}{C^2}, \tag{9}$$

for some cosmological constant Λ of $(p+D)$-dimensional space. The Einstein tensor's components from the metric are

$$G_0^0 = -\frac{p}{B^2}\frac{A''}{A} - \frac{p(p-1)}{2B^2}\left(\frac{A'}{A}\right)^2 + \frac{p}{B^2}\frac{A'B'}{AB} + \frac{(D-2)(D-3)}{2C^2}, \tag{10}$$

$$G_r^r = -\frac{p(p+1)}{2B^2}\left(\frac{A'}{A}\right)^2 + \frac{(D-2)(D-3)}{2C^2}, \tag{11}$$

$$G_\theta^\theta = -\frac{(p+1)}{B^2}\frac{A''}{A} - \frac{p(p+1)}{2B^2}\left(\frac{A'}{A}\right)^2 + \frac{(p+1)}{B^2}\frac{A'B'}{AB} + \frac{(D-4)(D-3)}{2C^2}. \tag{12}$$

Combining the above equations we have a useful expression

$$\begin{aligned} pG_\theta^\theta - (p+1)G_0^0 - G_r^r &= -2\Lambda - \frac{(p+D-2)\kappa\eta^2}{C^2} - \frac{(2p+D-2)(D-2)\kappa\eta^4}{2C^4\beta^2} \\ &= -\frac{(p+D-2)(D-3)}{C^2}. \end{aligned} \tag{13}$$

Solving this when $\Lambda = 0$ we get

$$C^2 = \frac{(2p+D-2)(D-2)\kappa\eta^4}{2\beta^2(p+D-2)(D-3-\kappa\eta^2)}, \tag{14}$$

and when $\Lambda \neq 0$ we get

$$\begin{aligned} C_\pm^2 = &\frac{(p+D-2)(D-3-\kappa\eta^2)}{4\Lambda} \\ &\pm\frac{\sqrt{(p+D-2)^2(D-3-\kappa\eta^2)^2 - 4\Lambda(2p+D-2)(D-2)\kappa\eta^4\beta^{-2}}}{4\Lambda}. \end{aligned} \tag{15}$$

The dimension is constrained with $p + D > 2$ since $D \geq 4$. We can see that (14) satisfies

$$\eta^2 < \eta_{crit}^2 \equiv (D-3)/\kappa \tag{16}$$

so that if (16) is applied to (14), $C_\pm^2$ is greater than zero. This condition also makes the first term in (15) greater than zero. As we limit (15) by $\beta \to \infty$, $C_-^2 \to 0$ and $C_+^2 \to (p+D-2)(D-3-\kappa\eta^2)/2\Lambda$. The latter is in agreement with [4].

Now we discuss metric solutions for the compactified space with manifold $Z_{p+2} \times S^{D-2}$. We obtain

$$B^{-2}\left(\frac{A'B'}{AB} - \frac{A''}{A}\right) = \pm\omega^2, \tag{17}$$

using $G_\theta^\theta - G_r^r$, and we define

$$\pm\omega^2 \equiv \frac{D-3}{(p+1)C^2} - \frac{(1 + 2\beta^{-2}(D-2)\eta^2/2C^2)}{(p+1)}\frac{\eta^2}{C^2}. \tag{18}$$

This can be considered as the cosmological constant of Z_{p+2} space for the following reasons. We use positive-negative signs to ensure $\omega^2 \geq 0$; the positive (negative) sign is for $B^{-2}\left(\frac{A'B'}{AB} - \frac{A''}{A}\right) > (<)\ 0$. Using $B = 1$, (17) gives us

$$ds^2 = \begin{cases} \frac{1}{\omega^2}(\sin^2\chi\ \eta_{\mu\nu}dx^\mu dx^\nu - d\chi^2) - C^2 d\Omega_{D-2}^2, & \text{for positive sign,} \\ \eta_{\mu\nu}dx^\mu dx^\nu - dr^2 - C^2 d\Omega_{D-2}^2, & \text{for } \omega = 0, \\ \frac{1}{\omega^2}(\sinh^2\chi\ \eta_{\mu\nu}dx^\mu dx^\nu - d\chi^2) - C^2 d\Omega_{D-2}^2, & \text{for negative sign,} \end{cases} \tag{19}$$

with $\chi \equiv \omega r$. Using $B = A^{-1}$,

$$ds^2 = \begin{cases} (1-\omega^2 r^2)\,\eta_{\mu\nu}dx^\mu dx^\nu - \frac{dr^2}{(1-\omega^2r^2)} - C^2 d\Omega^2_{D-2}, & \text{for positive sign,} \\ \eta_{\mu\nu}dx^\mu dx^\nu - dr^2 - C^2 d\Omega^2_{D-2}, & \text{for } \omega = 0, \\ (1+\omega^2 r^2)\,\eta_{\mu\nu}dx^\mu dx^\nu - \frac{dr^2}{(1+\omega^2r^2)} - C^2 d\Omega^2_{D-2}, & \text{for negative sign.} \end{cases} \tag{20}$$

These give us the manifold of compactified space:

$$Z_{p+2} = \begin{cases} dS_{p+2}, & \text{for } B^{-2}\left(\frac{A'B'}{AB} - \frac{A''}{A}\right) > 0, \\ M_{p+2}, & \text{for } B^{-2}\left(\frac{A'B'}{AB} - \frac{A''}{A}\right) = 0, \\ AdS_{p+2}, & \text{for } B^{-2}\left(\frac{A'B'}{AB} - \frac{A''}{A}\right) < 0. \end{cases} \tag{21}$$

3. Compactification solutions stability

To study the stability of above metric solutions (19) and (20) with their radius (14) or (15) depending on their cosmological constant Λ, we look for the effective potential of the radion $V(\psi)$ by dimensional reduction. Considering a metric with signature $(+ - - \ldots)$

$$ds^2 = G^{(p+D)}_{MN} dx^A dx^B, \tag{22}$$

with $G^{(p+D)}_{MN}$ metric on $(p+D)$-dimensional action

$$\mathcal{S} = \int d^{p+D}x\sqrt{|G|}\left[\frac{\mathcal{R}^{(p+D)}}{2\kappa} + \mathcal{L}_m\right]. \tag{23}$$

By defining the metric as

$$ds^2 = g^{(p+2)}_{\mu\nu}(x)dx^\mu dx^\nu - b^2(x)\gamma^{(D-2)}_{ij}(y)dy^i dy^j, \tag{24}$$

with $\gamma^{(D-2)}_{ij}$ a metric for a space with constant curvature with radius R_0, the action becomes

$$\mathcal{S} = \frac{V_{(D-2)}}{2\kappa}\int d^{p+2}x\sqrt{|g|}\left[b^{D-2}\mathcal{R}^{(p+D)} + 2\kappa b^{D-2}\mathcal{L}_m\right]. \tag{25}$$

Performing Weyl transformation to $\mathcal{R}^{(p+D)}$ by

$$g^{(p+2)}_{\mu\nu}(x) = b^{2a}(x)\,\tilde{g}^{(p+2)}_{\mu\nu}(x) \tag{26}$$

with $a = -(D-2)/p$, we obtain

$$\begin{aligned} \mathcal{S} \;=\; & \frac{V_{(D-2)}}{2\kappa}\int d^{p+2}x\sqrt{|\tilde{g}|}\left[\tilde{\mathcal{R}}^{(p+2)} + \frac{(D-2)(D-3)}{b^{2(p+D-2)/p}R_0^2} + \frac{2(D-2)}{p}\tilde{g}^{\mu\nu}\tilde{\nabla}_\mu(\tilde{\partial}_\nu \ln b)\right. \\ & \left. - \frac{(D-2)(p+D-2)}{p}\tilde{g}^{\mu\nu}(\tilde{\partial}_\mu \ln b)(\tilde{\partial}_\nu \ln b) + 2\kappa b^{-(D-2)/p}\tilde{\mathcal{L}}_m\right]. \end{aligned} \tag{27}$$

After we define the radion as

$$b \equiv \exp\left[\sqrt{\frac{p}{(D-2)(p+D-2)}}\frac{\psi}{M_P}\right], \tag{28}$$

with a $(p+2)$-dimensional Planck mass $M_P \equiv \sqrt{V_{(D-2)}/\kappa}$, we obtain an action in Einstein's frame

$$\mathcal{S} = \int d^{p+2}x\sqrt{|\tilde{g}|}\left[\frac{M_P^2\tilde{\mathcal{R}}^{(p+2)}}{2} - \frac{1}{2}\tilde{g}^{\mu\nu}\tilde{\partial}_\mu\psi\tilde{\partial}_\nu\psi - V(\psi)\right], \tag{29}$$

whose effective potential is

$$V(\psi) = -e^{\sigma\psi/M_P}\frac{M_P^2(D-2)(D-3)}{2R_0^2} - e^{\chi\psi/M_P}M_P^2\kappa\tilde{\mathcal{L}}_m, \tag{30}$$

with $\sigma \equiv -2\sqrt{(p+D-2)/p(D-2)}$ and $\chi \equiv -2\sqrt{(D-2)/p(p+D-2)}$; its metric is given by

$$ds^2 = e^{\chi\psi/M_P}\tilde{g}^{(p+2)}_{\mu\nu}dx^\mu dx^\nu - e^{\xi\psi/M_P}\gamma^{(D-2)}_{ij}dy^i dy^j, \tag{31}$$

with $\xi \equiv 2\sqrt{p/(D-2)(p+D-2)}$. We restrict $p \geq 1$ so that $\sigma \neq \infty$, $\chi \neq \infty$ and $\xi \neq 0$.

Now we can write metric (31) by setting it as

$$ds^2 = e^{\chi\psi/M_P}\left[A^2(r)\eta^{(p+1)}_{\mu\nu}dx^\mu dx^\nu - B^2(r)dr^2\right] - e^{\xi\psi/M_P}\left[C^2 d\Omega^2_{(D-2)}\right], \tag{32}$$

such that it resemble (1) as $\psi = 0$. The radius R_0 now becomes C. By applying (32) into the matter terms in (3) we obtain the Lagrangian density $\tilde{\mathcal{L}}_m$ in (30) as

$$\tilde{\mathcal{L}}_m = -\frac{\Lambda}{\kappa} + K(X), \quad \text{with } X = \frac{(D-2)\eta^2}{2C^2}e^{-\xi\psi/M_P}, \tag{33}$$

with $K(X) = -X - \beta^{-2}X^2$. This hence give us the effective potential for our model

$$\begin{aligned} V(\psi) &= -e^{\sigma\psi/M_P}\frac{M_P^2(D-2)(D-3)}{2C^2} + e^{\chi\psi/M_P}M_P^2\Lambda \\ &\quad + M_P^2\kappa\left[\frac{(D-2)\eta^2}{2C^2}e^{(\chi-\xi)\psi/M_P} + \frac{(D-2)^2\eta^4}{4\beta^2C^4}e^{(\chi-2\xi)\psi/M_P}\right]. \end{aligned} \tag{34}$$

Choosing the extrema to be at $\psi = 0$, we solve $V'(\psi = 0) = 0$ or

$$0 = \chi\Lambda + \frac{(D-2)}{2}\left[(\chi-\xi)\kappa\eta^2 - \sigma(D-3)\right]\frac{1}{C^2} + \left[\frac{(\chi-2\xi)(D-2)^2\kappa\eta^4}{4\beta^2}\right]\frac{1}{C^4}, \tag{35}$$

then we obtain

$$C^2 = \frac{(2p+D-2)(D-2)\kappa\eta^4}{2\beta^2(p+D-2)(D-3-\kappa\eta^2)}, \tag{36}$$

when $\Lambda = 0$, and

$$C^2 = \frac{(p+D-2)(D-3-\kappa\eta^2)}{4\Lambda}\left[1 \pm \sqrt{1 - \frac{(2p+D-2)(D-2)\kappa\eta^4\beta^{-2}}{4\Lambda(p+D-2)^2(D-3-\kappa\eta^2)^2}}\right], \tag{37}$$

when $\Lambda \neq 0$. Equation (37) needs to be $C^2 = (p+D-2)(D-3-\kappa\eta^2)/2\Lambda$ at $\beta \to \infty$, which is only satisfied with positive sign in (37), not the minus sign. Both are the same as (14) and (15), which we find using Einstein equations. By choosing the positive sign, it is clear that Λ cannot have negative value, otherwise it makes $C^2 < 0$. Both solutions have a stability condition

$$\begin{aligned} V''(\psi = 0) &= \frac{4(D-2)\Lambda}{p(p+D-2)} + \frac{2(\kappa\eta^2 - D + 3)(p+D-2)}{pC^2} + \frac{(2p+D-2)^2(D-2)\kappa\eta^4}{p(p+D-2)\beta^2C^4} \\ &> 0. \end{aligned} \tag{38}$$

Equations (36) and (37) have a more compact expression with arbitrary Λ

$$\frac{1}{C_\pm^2} = \frac{(D-3-\kappa\eta^2)(p+D-2)\beta^2}{(2p+D-2)(D-2)\kappa\eta^4}\left(1 \pm \sqrt{1 - \frac{4\Lambda(2p+D-2)(D-2)\kappa\eta^4}{(p+D-2)^2(D-3-\kappa\eta^2)^2\beta^2}}\right), \tag{39}$$

which arises from (35). We use C_+ to ensure it becomes (36) when $\Lambda = 0$, but we use C_- for $\Lambda \neq 0$ due to $C_-^2 = (p+D-2)(D-3-\kappa\eta^2)/2\Lambda$ at $\beta \to \infty$ thus C_- corresponds to the positive sign in (37). The term under the square root needs to be positive, and thus we need $\beta^2 > 0$ when $\Lambda \leq 0$ or

$$\beta^2 \geq \beta_{crit}^2 \equiv \frac{4\Lambda(2p+D-2)(D-2)\kappa\eta^4}{(p+D-2)^2(D-3-\kappa\eta^2)^2}, \tag{40}$$

when $\Lambda > 0$. It is clear that (16) must be satisfied. We set the minimum of the potential positioned at $\psi = 0$ for simplicity.

When the above condition is satisfied, we can set $V(\psi = 0, \Lambda = \Lambda_*) = 0$. Substituting (39) in to (35) to get

$$\begin{aligned} 0 &= \Lambda_* - \frac{(D-3-\kappa\eta^2)^2 p(p+D-2)\beta^2}{2(2p+D-2)^2\kappa\eta^4}\left(1 + \frac{2\Lambda_*(2p+D-2)(D-2)\kappa\eta^4}{p(p+D-2)(D-3-\kappa\eta^2)^2\beta^2}\right) \\ &\mp \frac{(D-3-\kappa\eta^2)^2 p(p+D-2)\beta^2}{2(2p+D-2)^2\kappa\eta^4}\sqrt{1 - \frac{4\Lambda_*(2p+D-2)(D-2)\kappa\eta^4}{(p+D-2)^2(D-3-\kappa\eta^2)^2\beta^2}}. \end{aligned} \tag{41}$$

Since the left hand side is equal to zero, we can arrange it to get

$$\pm\sqrt{1 - \frac{4\Lambda_*(2p+D-2)(D-2)\kappa\eta^4}{(p+D-2)^2(D-3-\kappa\eta^2)^2\beta^2}} = \frac{2\Lambda_*(2p+D-2)(2p)\kappa\eta^4}{p(p+D-2)(D-3-\kappa\eta^2)^2\beta^2} - 1. \tag{42}$$

After squaring both sides we get $\Lambda_* = 0$ and

$$\Lambda_* = \frac{(D-3-\kappa\eta^2)^2\beta^2}{4\kappa\eta^4}. \tag{43}$$

We will use the latter expression and we restrict $\Lambda_* > 0$ so that (16) is satisfied. This treatment follows the method used in [16].

To see whether this compactification is stable we need to investigate the second derivative of the potential. After substituting (43) into (39), the inverse of radius squared becomes

$$\frac{1}{C_\pm^2} = \frac{4(p+D-2)\Lambda_*}{(2p+D-2)(D-2)(D-3-\kappa\eta^2)}\left(1 \pm \sqrt{1 - \frac{\Lambda(2p+D-2)(D-2)}{\Lambda_*(p+D-2)^2}}\right). \tag{44}$$

This expression tells us that Λ cannot have higher value than $\Lambda = (p+D-2)^2\Lambda_*/(2p+D-2)(D-2)$ otherwise $1/C_\pm^2$ will be complex valued. Substituting (44) into (38) gives us

$$\begin{aligned} V''(\psi = 0, C_\pm^2) &= \frac{2(p+D-2)(D-3-\kappa\eta^2)^2\beta^2}{(2p+D-2)(D-2)\kappa\eta^4}\left(1 \pm \sqrt{1 - \frac{4\Lambda(2p+D-2)(D-2)\kappa\eta^4}{(p+D-2)^2(D-3-\kappa\eta^2)^2\beta^2}}\right) \\ &\quad - \frac{8\Lambda}{(p+D-2)} \\ &= -\frac{8\Lambda}{(p+D-2)} + \frac{8(p+D-2)\Lambda_*}{(2p+D-2)(D-2)}\left(1 \pm \sqrt{1 - \frac{\Lambda(2p+D-2)(D-2)}{(p+D-2)^2\Lambda_*}}\right) \end{aligned} \tag{45}$$

where positive and negative signs correspond to C_+ and C_-, respectively. If $V'' \geq 0$ then the minimum of the potential at $\psi = 0$ can be determined by

$$\begin{aligned} \frac{V(\psi=0, C_\pm^2)}{M_P^2} &= \frac{2p\Lambda}{(2p+D-2)} \\ &\quad - \frac{p(p+D-2)(D-3-\kappa\eta^2)^2\beta^2}{(2p+D-2)^2 2\kappa\eta^4}\left(1 \pm \sqrt{1 - \frac{4\Lambda(2p+D-2)(D-2)\kappa\eta^4}{(p+D-2)^2(D-3-\kappa\eta^2)^2\beta^2}}\right) \\ &= \frac{2p\Lambda}{(2p+D-2)} - \frac{2p(p+D-2)\Lambda_*}{(2p+D-2)^2}\left(1 \pm \sqrt{1 - \frac{\Lambda(2p+D-2)(D-2)}{(p+D-2)^2\Lambda_*}}\right). \end{aligned} \tag{46}$$

Now we can determine the compactified space and its stability. In the case of $\Lambda = 0$,

$$\frac{V(\psi=0, C_+, \Lambda=0)}{M_P^2} = -\frac{4p(p+D-2)\Lambda_*}{(2p+D-2)^2} < 0, \tag{47}$$

$$V''(\psi=0, C_+, \Lambda=0) = \frac{16(p+D-2)\Lambda_*}{(2p+D-2)(D-2)} > 0, \tag{48}$$

since $\Lambda_* > 0$ which imply that the compactification channel is $M_{p+D} \to AdS_{p+2} \times S^{D-2}$ and stable. In the case of $\Lambda > 0$,

$$\frac{V(\psi=0, C_-)}{M_P^2} = \frac{2p\Lambda}{(2p+D-2)} - \frac{2p(p+D-2)\Lambda_*}{(2p+D-2)^2}\left(1 - \sqrt{1 - \frac{\Lambda(2p+D-2)(D-2)}{(p+D-2)^2\Lambda_*}}\right), \tag{49}$$

$$V''(\psi=0, C_-) = -\frac{8\Lambda}{(p+D-2)} + \frac{8(p+D-2)\Lambda_*}{(2p+D-2)(D-2)}\left(1 - \sqrt{1 - \frac{\Lambda(2p+D-2)(D-2)}{(p+D-2)^2\Lambda_*}}\right). \tag{50}$$

Now we consider positive but nearly zero Λ

$$\frac{V(\psi=0, C_-)}{M_P^2} \simeq \frac{2p\Lambda}{(2p+D-2)} - \frac{\Lambda(D-2)p}{(2p+D-2)(p+D-2)} = \frac{2p\Lambda}{(p+D-2)} \simeq 0, \tag{51}$$

$$V''(\psi=0, C_-) \simeq -\frac{8\Lambda}{(p+D-2)} + \frac{8\Lambda}{(p+D-2)} = 0, \tag{52}$$

and at $\Lambda = (p+D-2)^2\Lambda_*/(2p+D-2)(D-2)$

$$\frac{V(\psi=0, C_-)}{M_P^2} = \frac{2p(p+D-2)^2\Lambda_*}{(2p+D-2)^2(D-2)} - \frac{2p(p+D-2)\Lambda_*}{(2p+D-2)^2} = \frac{2p^2(p+D-2)\Lambda_*}{(2p+D-2)^2(D-2)}, \tag{53}$$

$$V''(\psi=0, C_-) = -\frac{8(p+D-2)\Lambda_*}{(2p+D-2)(D-2)} + \frac{8(p+D-2)\Lambda_*}{(2p+D-2)(D-2)} = 0. \tag{54}$$

Both cases need the third derivative of V, which is

$$V'''(\psi=0, C_-) \simeq 8\sqrt{\frac{p}{(D-2)(p+D-2)}}\frac{\Lambda_*}{M_P}\frac{\left[(p+D-2)^3 - (D-2)^2\right]}{p^2(p+D-2)} > 0 \tag{55}$$

for $\Lambda \ll \Lambda_*$ case, and

$$V'''(\psi=0, C_-) = \frac{\left[(p+D-2)^3 - (D-2)^2\right]}{(D-2)p^2}\frac{\Lambda_*}{M_P}\sqrt{\frac{p}{(D-2)(p+D-2)}} > 0 \tag{56}$$

for $\Lambda = (p + D - 2)^2\Lambda_*/(2p + D - 2)(D - 2)$ case, respectively. These imply that the compactification channels $dS_{p+D} \rightarrow (dS, M)_{p+2} \times S^{D-2}$ are unstable since the potential at $\psi = 0$ is a strictly increasing point of inflection. Combining these results we get tunneling channels $Y_{p+D} \longrightarrow Z_{p+2} \times S^{D-2}$ (with Λ and $V(\psi = 0)$ denoting Y_{p+D} and Z_{p+2} respectively)

$$dS_{p+D} \longrightarrow \begin{cases} dS_{p+2} \times S^{D-2}, \\ M_{p+2} \times S^{D-2}, \end{cases} \tag{57}$$

$$M_{p+D} \longrightarrow AdS_{p+2} \times S^{D-2}. \tag{58}$$

4. Conclusions

Here we investigate compactification by k-monopole in nonlinear sigma model with kinetic terms $K(X) = -X - \beta^{-2}X^2$. We can use both Einstein equations or effective potential from dimensional reduction to determine radius of compactification. The compactification channels can be found by introducing effective potential and its derivatives will determine the stability. Here we report that the compactification channels from M_{p+D} and dS_{p+D} to $Z_{p+2} \times S^{D-2}$, where Z is some topological manifold, are stable and unstable, respectively.

Acknowledgments

IP thanks Muhammad Iqbal and Brian Agung Cahyo for the discussions on compactification. This work is partially funded by Universitas Indonesia.

References

[1] Kibble T W B 1976 *J. Phys.* A **9** 1387
[2] Shellard E and Vilenkin A 1994 *Cosmic Strings and Other Topological Defects* (Cambridge: Cambridge University Press) p 64
[3] Barriola M and Vilenkin A 1989 *Phys. Rev. Lett.* **63** 341
[4] Olasagasti I and Vilenkin A 2000 *Phys. Rev.* D **62** 044014
[5] Cho I and Vilenkin A 1997 *Phys. Rev.* D **56** 7621
[6] Cho I and Vilenkin A 2003 *Phys. Rev.* D **68** 025013
[7] Liebling S L 2000 *Phys. Rev.* D **61** 024030
[8] Vilenkin A 1994 *Phys. Rev. Lett.* **72** 3137
[9] Linde A D 1994 *Phys. Lett.* B **327** 208
[10] Skyrme T H R 1961 *Proc. R. Soc. Lond.* A **262** 237
[11] Babichev E 2006 *Phys. Rev.* D **74** 085004
[12] Jin X h, Li X z and Liu D j 2007 *Class. Quant. Grav.* **24** 2773
[13] Li X z and Liu D j 2005 *Int. J. Mod. Phys.* A **20** 5491
[14] Liu D J, Zhang Y L and Li X Z 2009 *Eur. Phys. J.* C **60** 495
[15] Prasetyo I and Ramadhan H S 2016 *Gen. Rel. Grav.* **48** 10
[16] Blanco-Pillado J J, Schwartz-Perlov D and Vilenkin A 2009 *J. Cosmol. Astropart. Phys.* JCAP12(2009)006
[17] Gell-Mann M and Zwiebach B 1984 *Phys. Lett.* B **141** 333

IOP Conf. Series: Journal of Physics: Conf. Series **856** (2017) 012009 doi:10.1088/1742-6596/856/1/012009

First-order perturbative approach to Q-balls with massive gauge field

M F A R Sakti[1], P Y D Sagita[1], A Suroso[1,2], A Sulaksono[3] and F P Zen[1,2]

[1] Theoretical Physics Division, Institut Teknologi Bandung, Jl. Ganesha 10, Bandung 40132, Indonesia

[2] Indonesia Center for Theoretical and Mathematical Physics (ICTMP), Jl. Ganesha 10, Bandung 40132, Indonesia

[3] Department of Physics, Universitas Indonesia, Depok 16424, Indonesia

E-mail: m.fitrah@students.itb.ac.id

Abstract. In this paper, we examine the properties of massive $U(1)$ gauged Q-balls. For finding the field solutions, we use first-order perturbative approach where the coupling constant q is used as pertubation parameter. We also compare our result to the massless case to see the effects of the gauge field mass. Finally, we obtain the mass-charge relation of the Q-balls.

1. Introduction

Nontopological solitons [1] or Q-balls [2] represent stationary localized solutions in flat space possessing a finite mass. In contrast to topological solitons, nontopological solitons are stable not because of topological charge, but due to a globally conserved Noether charge that comes from a continuous symmetry existing in the system. In the simplest case of nontopological solitons, Q-balls consist of complex scalar field that has suitable self-interaction. The global phase invariance of scalar field theory is connected to Noether charge, symbolized by Q, corresponding to particle number of the Q-balls. Generally when the gravity is invoked, the Q-balls solutions will be boson stars (BSs) solutions [3–5]. The scalar field on Q-balls actually can be a model to describe dark matter [7] or on BSs to describe dark energy (DE) [6, 14] .

These Q-balls solutions appear with possessing a global continuous $U(1)$ symmetry. When the $U(1)$ symmetry is local or gauged, the product between Noether charge and the constant defining the coupling between scalar field and the gauge one can be interpreted as the physical charge of the solitons [4, 8] but for convenience we refer to [1, 5, 9, 10] that there is no product with coupling constant to obtain the charge. For the non-gauged Q-balls, the solutions exist only in certain frequency range, i.e., $\omega_{min} < \omega < \omega_{max}$ determined by the properties of the potential [1]. At the highest critical value of the frequency, both mass and charge of the Q-balls are assumed in their maximum values. If we go across the limit of ω, indeed, we will see that the solutions have some nodes.

It is clear that, for the small gauge field on the scalar field, the properties of these Q-balls solutions will not differ considerably from the non-gauged Q-balls. So, the interesting case happens if the gauge field considerably affects the properties of the Q-balls. In general case, the gauge field can possess mass hence the gauge invariance is not satisfied [11]. In this paper, we examine gauged Q-balls solutions with no self-interacting term in potential because we assume the interaction that holds the Q-balls to be stable comes from interaction between scalar field and gauge field. In addition, we add the term of mass of the gauge field. We find the solutions by using perturbative approach where the coupling constant q is used as a perturbative parameter, and we just examine to the first-order solutions. Next, we find the mass-charge relation for this solution and compare it with the massless case.

The paper is organized as follows. In Section 2, we present the general setup and introduce the notations that will be used throughout the paper. In Section 3, we present analytically perturbative approach results of the field solutions related to (massive) gauged Q-balls. In Section 4, we examine the solutions, pressures, mass and charge of the Q-balls. The results are discussed in the last section.

2. General setup and Lagrangian

We describe Q-balls that consist of scalar field (Φ) and gauge field (A_α) with the Lagrangian density, in natural units with $\hbar = c = 1$, as

$$\mathcal{L} = g^{\alpha\beta}(D_\alpha\Phi)^*(D_\beta\Phi) - {m_b}^2\Phi^*\Phi - \frac{1}{4}F_{\alpha\beta}F^{\alpha\beta} + \frac{1}{2}\mu^2 A_\alpha A^\alpha. \tag{1}$$

Here m_b is the scalar field mass, μ is the gauge field mass, $D_\alpha = \partial_\alpha + iqA_\alpha$, where q is coupling constant of the interaction strength and $F_{\alpha\beta} = \partial_\alpha A_\beta - \partial_\beta A_\alpha$. Then we set stationary scalar field ansatz [2]

$$\Phi(r,t) = \phi(r)e^{i\omega t}. \tag{2}$$

Equation (2) describes a spherically symmetric bound state of scalar fields with eigenfrequency ω. Accordingly, we set the gauge field to having only time-component, so we have

$$A_\alpha dx^\alpha = A(r)dt. \tag{3}$$

In contrast to the electromagnetic field with no coupling, which is gauge invariant, our massive gauge field will not satisfy any gauge conditions [11]. We do the analytic calculations to obtain some differential equations of motion related to Langrangian (1), as done in [3,12]. The conserved current can be defined in the following

$$J_\alpha = \left[i(\Phi^*\partial_\alpha\Phi - \Phi\partial_\alpha\Phi^*) - 2qA_\alpha\Phi^*\Phi\right], \tag{4}$$

so energy-momentum tensor T^α_β becomes

$$T^\alpha_\beta = g^{\sigma\alpha}\left[(D_\sigma\Phi)^*(D_\beta\Phi) + (D_\beta\Phi)^*(D_\sigma\Phi) - g^{\alpha\beta}F_{\alpha\sigma}F_{\beta\rho} + \mu^2 A_\sigma A_\beta\right] - \delta^\alpha_\beta\mathcal{L}. \tag{5}$$

From (4), we obtain the number density defined by

$$J_0 = 2(\omega + qA)\phi^2. \tag{6}$$

By using Euler-Lagrange equation, we obtain the system of equations for scalar and gauge fields, i.e.,

$$\phi'' + \frac{2}{r}\phi' + (\omega + qA)^2\phi - \frac{1}{2}\frac{dV(\phi)}{d\phi} = 0, \tag{7}$$

$$A'' + \frac{2}{r}A' - 2q\,(\omega + qA)\,\phi^2 - \mu^2 A = 0, \tag{8}$$

IOP Conf. Series: Journal of Physics: Conf. Series **856** (2017) 012009 doi:10.1088/1742-6596/856/1/012009

where we choose $V(\phi) = m_b^2\phi^2$. Next, the energy density ρ is defined by

$$\rho = T_0^0 = \left[m_b^2 + (\omega + qA)^2\right]\phi^2 + \frac{A'^2}{2} + \phi'^2 + \frac{1}{2}\mu^2 A^2, \tag{9}$$

and the other nonvanishing components are

$$T_1^1 = \left[m_b^2 - (\omega + qA)^2\right]\phi^2 + \frac{A'^2}{2} - \phi'^2 - \frac{1}{2}\mu^2 A^2, \tag{10}$$

$$T_2^2 = T_3^3 = \left[m_b^2 - (\omega + qA)^2\right]\phi^2 - \frac{A'^2}{2} + \phi'^2 - \frac{1}{2}\mu^2 A^2, \tag{11}$$

which correspond to the pressures of the Q-balls in directions of r and θ, respectively. The charge of the Q-balls can be defined by [1, 5, 9, 10]

$$Q = 8\pi \int_0^\infty (\omega + qA)\phi^2 r^2 dr, \tag{12}$$

which is also physically defined as the particle number of the Q-balls. In addition, the physical charge for gauged Q-balls is $\overline{Q} = qQ$ [4, 8]. The mass can also be given by

$$M = 4\pi \int_0^\infty m_b^2\phi^2 + (\omega + qA)^2\phi^2 + \frac{A'^2}{2} + \phi'^2 + \frac{1}{2}\mu^2 A^2\Big) r^2 dr. \tag{13}$$

Next, fields will be given in the following rescaling parameters

$$\phi \to \phi m_b, \quad A \to A m_b, \quad \omega \to \omega m_b, \quad \mu \to \mu m_b, \quad r \to r/m_b. \tag{14}$$

3. Perturbative approximation for the fields

We use perturbative approach to obtain the solutions of KGP (also for KGM). First, we will assume that q is a small constant so it will be our perturbative parameter. So, the expansions of the fields become

$$\phi = \phi_0 + q\phi_1 + q^2\phi_2 + ... \tag{15}$$

$$A = A_0 + qA_1 + q^2 A_2 + ... \tag{16}$$

Then by inserting (15) and (16) to (7) and (8), we will obtain

$$\begin{aligned}&(\phi_0'' + q\phi_1'' + ...) + \frac{2}{r}(\phi_0' + q\phi_1' + ...) - m^2(\phi_0 + q\phi_1 + ...) \\ &= -2q\omega(A_0 + qA_1 + ...)(\phi_0 + q\phi_1 + ...) - q^2(A_0 + qA_1 + ...)^2(\phi_0 + q\phi_1 + ...),\end{aligned} \tag{17}$$

$$\begin{aligned}&(A_0'' + qA_1'' + ...) + \frac{2}{r}(A_0' + qA_1' + ...) - \mu^2(A_0 + qA_1 + ...) \\ &= 2q\omega(\phi_0 + q\phi_1 + ...)^2 + 2q^2(A_0 + qA_1 + ...)(\phi_0 + q\phi_1 + ...)^2,\end{aligned} \tag{18}$$

where $\omega^2 - 1 = -m^2$. We take from (17) and (18) to the first-order. So we will have

$$\phi_0'' + \frac{2}{r}\phi_0' - m^2\phi_0 = 0, \tag{19}$$

$$A_0'' + \frac{2}{r}A_0' - \mu^2 A_0 = 0, \tag{20}$$

for the zeroth-order and

$$\phi_1'' + \frac{2}{r}\phi_1' - m^2\phi_1 = -2\omega A_0\phi_0, \tag{21}$$

$$A_1'' + \frac{2}{r}A_1' - \mu^2 b = 2\omega\phi_0^2, \tag{22}$$

for the first-order. To obtain the arbitrary solutions of KGP and KGM equations, we need the boundary conditions that shows a localized particle distribution and remove the singularity on $r = 0$. We impose the following regular boundary conditions:

$$\phi(r \to \infty) = 0, \quad \phi'(r = 0) = 0. \tag{23}$$

We also impose the gauge field to be monotonically decreasing so that

$$A(r \to \infty) = 0, \quad A'(r = 0) = 0. \tag{24}$$

By applying (23) and (24) to (19) - (22), we obtain the solutions for the first-order KGP, they are

$$\phi(r) = \frac{ae^{-mr}}{r} + \frac{qe^{-mr}}{r}\left[c + \frac{ab\omega}{m}\left(Ei(-\mu r) - e^{2mr}Ei[-(2m+\mu)r]\right)\right], \tag{25}$$

and

$$A(r) = \frac{be^{-\mu r}}{r} + \frac{qe^{-\mu r}}{r}\left[d - \frac{a^2\omega}{\mu}\left(Ei[(-2m+\mu)r] - e^{2\mu r}Ei[-(2m+\mu)r]\right)\right], \tag{26}$$

where $Ei(-at) = \int_t^\infty e^{-at}/t\, dt$ and a, b, c, d are integration constants. For this case, there is an upper limit for the frequency ω that leads to the following condition [8]

$$\omega^2 < \frac{1}{2}V''(0) = 1. \tag{27}$$

For Maxwell case, the solutions are

$$\phi(r) = \frac{ae^{-mr}}{r} + \frac{qe^{-mr}}{r}\left[c + \frac{ab\omega}{m}\left(ln[r] - e^{2mr}Ei[-2mr]\right)\right], \tag{28}$$

and

$$A(r) = \frac{b}{r} + \frac{q}{r}\left[d + \frac{a^2\omega}{m}\left(2mrEi[-2mr] + e^{-2mr}\right)\right]. \tag{29}$$

4. Q-balls solutions

We show plots of the field profiles in figures 1 and 2 with ω in the lowest limit, near upper limit and between them. We also set constant values of q and μ. Figure 1 shows that the different values of ω make the scalar field change, with larger ω making larger $\phi(r)$. However, in Proca case, the near upper limit is dominant as shown in the figure. Scalar field profile for Maxwell case is also bigger than the Proca (massive gauge field) case. The initial value of the scalar field, $\phi(0)$, can be determined by setting the values of a, b, c and d. Figure 2 shows the gauge field profiles where the configurations are same with the scalar fields. The properties of the gauge fields are increasing simultaneously with the addition of the ω. Indeed, the increasing of the gauge mass μ makes the solutions smaller. It is because the gauge mass behaves like a damping

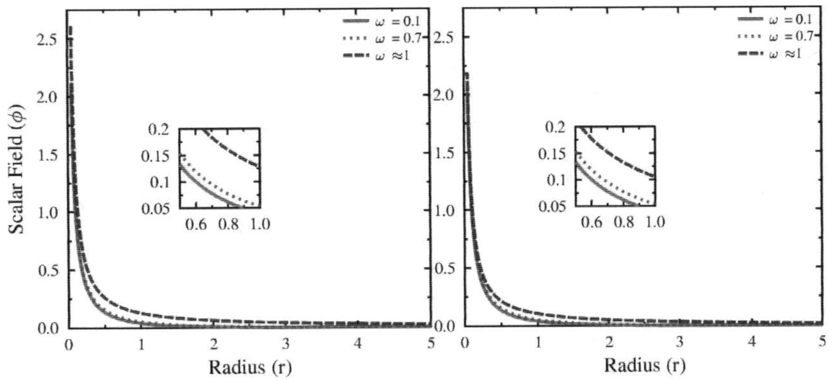

Figure 1. Scalar field profile for Maxwell (left) and Proca (right) with three different values of ω.

factor that decreases the value of the field. Again, we can set the initial value of $A(r)$, i.e., $A(0)$ with the same constants.

The pressures of the Q-balls are shown in figures 3 and 4. Figure 3 shows the radial pressure of the Q-balls obtained by inserting the solutions (25)–(29) to (10). Radial pressure approaches zero at large r, which shows that the Q-balls have a finite mass. From figure 3, we can see that there is no significant effect of the gauge mass μ and ω for the radial pressure in all range of radius. However, different values of ω affect the initial value of the pressure because it has significant initial value.

Figure 4 shows the pressure in tangential direction. This pressure has similar properties with the radial pressure but the initial values are different. Tangential pressure has negative values. Mathematically, this negativeness is caused by the value of the scalar field derivatives (that have negative sign) that are larger than the fields itself. Possibly, the physical meaning of this negativeness is the existence of DE in the Q-balls like in the dark-energy star [14] but just only in the tangential pressure. In the case of Q-balls, we also consider that there is no gravitational term such that the Lagrangian of the matter does not couple to the gravity. In our universe, DE makes it accelerate and develope because the pressures are negative [6, 16]. DE also can be modeled by scalar fields with arbitrary potential; hence it yields the negative pressure and equation of state $w_\phi = P_\phi/\rho_\phi < -1/3$, such in the Quintessence model [16]. In our case of

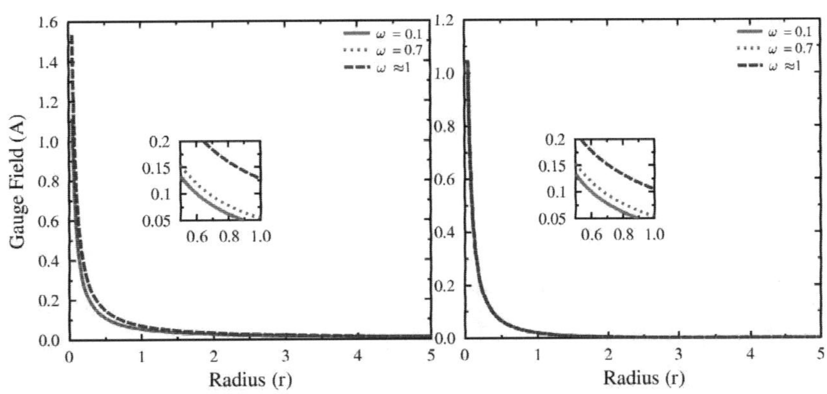

Figure 2. Gauge field profile for Maxwell (left) and Proca (right) with three different values of ω.

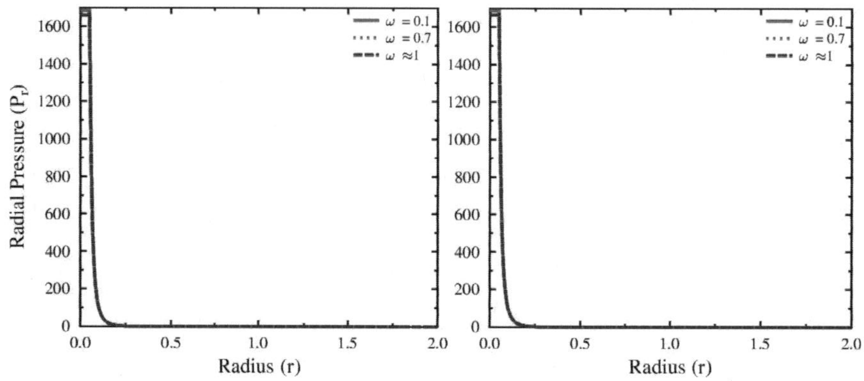

Figure 3. Radial pressure for Maxwell (left) and Proca (right) with three different values of ω.

Q-balls, we obtain the equation of state $w_t = P_t/\rho \simeq -0.76$, which satisfies the DE condition. To find the value of w_t, we take the sum of tangential pressure from every point given in figure 4. We then divide it by the number of points of the tangential pressure. We do the similar way for the energy density. As the result, we get the the average values of the tangential pressure and the energy density, which we can use to obtain the value of w_t. Generally, our Q-balls are suitable with all energy conditions [14, 15] where the more detailed discussion are prepared for another manuscript.

We step to the other characteristics of the Q-balls, i.e., mass and charge. Mass M and charge Q can be easily integrated from (12) and (13). These are shown in figure 5. Mass profile $M(r)$ are shown in the left panel of figure 5 with three different values of ω for Maxwell and Proca cases. As we see from mass-profile figure, charge profile $Q(r)$ of Q-balls has similar properties where its values are increasing with the addition of the radius of Q-balls, see figure 5 (middle). The values also increase when ω and μ increase. The relation between M and Q can be obtained qualitatively from the right panel of figure 5. Every reference gives their argument about the relation between mass (or energy) end charge. As discussed in [17], it must satisfy linear relation

$$M = \omega Q, \tag{30}$$

for gauged Q-balls that is similar with ordinary Q-balls [5]. In [10] for non-massive gauge the

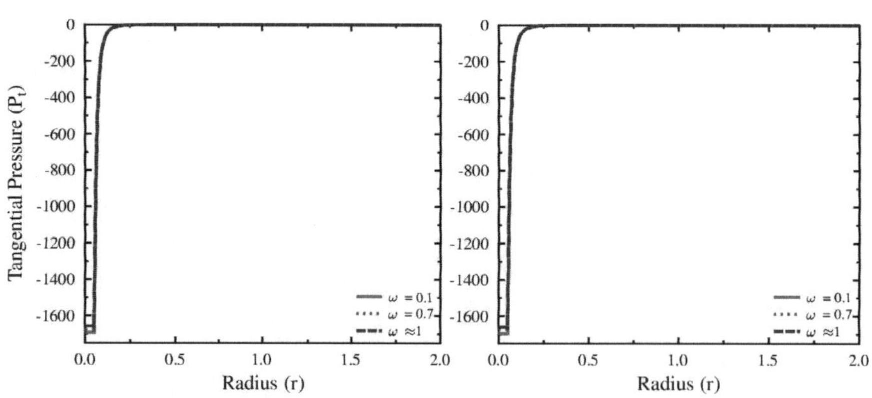

Figure 4. Tangential pressure for Maxwell (left) and Proca (right) with three different values of ω.

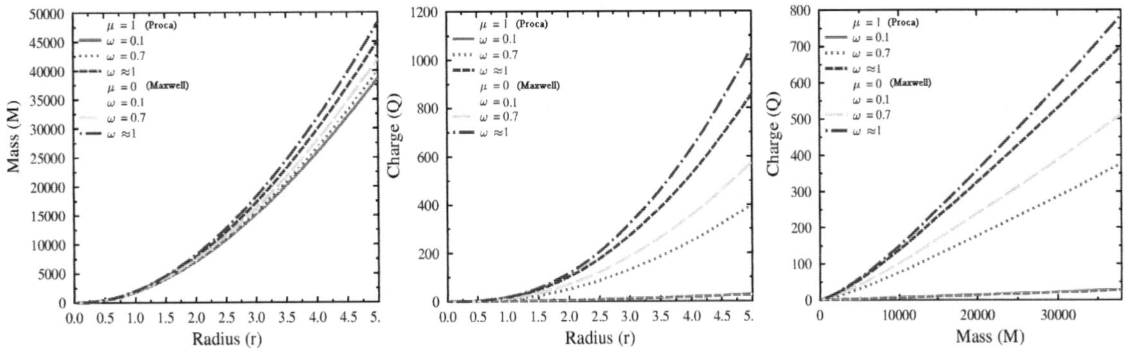

Figure 5. Profiles of mass M (left), charge Q (middle), and mass-charge relation M–Q (right) with three different values of ω for Maxwell and Proca cases.

relation becomes

$$M = \omega Q + \frac{4\pi}{3}\int_0^\infty drr^2\left[\phi'(r)^2 + q^2 A'(r)^2\right]. \tag{31}$$

But (31) will not reduce to ordinary Q-balls when $A = 0$. The detailed description about this relation in our case will be explained in another manuscript. In our qualitative observation, for our Q-balls (Maxwell and Proca), the condition for the relation between M and Q is almost linear and satisfies the condition (31) where the value of ω is limited under the value of m_b. But we still have to calculate the second term in right-hand side of (31) that plays a role as (may be) nonlinear term.

5. Conclusions

As we have demonstrated above, we have addressed gauged Q-balls with massless gauged field (Maxwell) and massive gauged field (Proca). We find the solution by using perturbative approach to the first-order. Field profiles from those two cases have similar properties such as we can see from the variation of position r and ω. As the consequence of the field solutions, the pressure, mass and charge also have similar properties. In our model of Q-balls, we also find that the tangential pressure behaves like DE model of scalar fields. Qualitatively, we obtain that M and Q have a linear relation such in some references where the value of ω is limited under the value of m_b.

Acknowledgments

We gratefully acknowledge support by Riset PMDSU 2016 from Ministry of Research, Technology, and Higher Education of the Republic of Indonesia. M.F.A.R.S would like to thank all members of Theoretical Physics Laboratory, Institut Teknologi Bandung for the kind discussion.

References

[1] Lee T D and Pang Y 1992 *Phys. Rep.* **221** 251
[2] Coleman S 1985 *Nuc. Phys.* B **262** 263
[3] Sakti M F A R and Sulaksono A 2016 *J. Phys.: Conf. Ser.* **739** 012075
[4] Pugliese D, Quevedo H, Rueda J A and Ruffini R 2013 *Phys. Rev.* D **88** 024053
[5] Kleihaus B, Kunz J and List M 2013 *Phys. Rev.* D **72** 064002
[6] Zen F P, Arianto, Gunara B E, Triyanta and Purwanto A 2009 *Eur. Phys. J.* C **63** 477
[7] Kusenko A and Shaposhnikov M 1998 *Phys. Lett.* B **418** 46
[8] Brihaye Y, Diemer V and Hartmann B 2014 *Phys. Rev.* D **89** 084048

[9] Hong J P, Kawasaki M and Yamada M 2015 *Phys. Rev.* D **92** 063521
[10] Lee K, Stein-Schabes J A, Watkins R and Widrow L M 1989 *Phys. Rev.* D **39** 1665
[11] Greiner W and Reinhardt J 1996 *Field Quantization* (Berlin: Springer-Verlag)
[12] Sakti M F A R and Sulaksono A 2016 *AIP Conf. Proc.* **1729** 020035
[13] Tamaki T and Sakai N 2014 *Phys. Rev.* D **90** 085022
[14] Horvat D and Marunović A 2013 *Class. Quantum Grav.* **30** 145006
[15] Rahaman F, Maulick R, Yadav A K, Ray S and Sharma R 2012 *Gen. Relativ. Gravit.* **44** 107
[16] Amendola L and Tsujikawa S 2010 *Dark Energy: Theory and Observations* (Cambridge: Cambridge University Press)
[17] Gulamov I E, Nugaev E Y and Smolyakov M N 2014 *Phys. Rev.* D **89** 085006

IOP Conf. Series: Journal of Physics: Conf. Series **856** (2017) 012010 doi:10.1088/1742-6596/856/1/012010

Variational iteration method used to solve the one-dimensional acoustic equations

Paskalia Siwi Setianingrum[1] **and Sudi Mungkasi**[2]

[1] Postgraduate Program in Mathematics Education, Faculty of Teacher Training and Education, Sanata Dharma University, Mrican, Tromol Pos 29, Yogyakarta 55002, Indonesia
[2] Department of Mathematics, Faculty of Science and Technology, Sanata Dharma University, Mrican, Tromol Pos 29, Yogyakarta 55002, Indonesia

E-mail: paskaliasiwi@yahoo.com, sudi@usd.ac.id

Abstract. Variational iteration method has been successful to solve a wide range of linear and nonlinear problems. This method is convergent to the exact solution, and furthermore, if there is an explicit form of the exact solution, then the method converges very rapidly. In this paper, we consider acoustic equations. Our contribution is a new application of the variational iteration method for solving the acoustic equations. Results show that the variational iteration method gives physically correct approximate solutions to the acoustic equations. We demonstrate the method solving the acoustic equations using several iterations.

1. Introduction

A mathematical model needs to be solved to find the solution to a real problem. It is generally difficult to find analytical exact solutions to mathematical models if the models contain many variables and have complex structures. In this case, an approximate approach gives a way to find solutions to real problems.

Elasticity equations are a model for wave propagation. Elasticity equations can be simplified into acoustic equation under some assumptions. Acoustics include sound waves. Acoustic equations model pressure and velocity changes of a system. One application of acoustic equations is the vibration of a cable wire that causes sound waves.

One of methods that can solve the acoustic equations is the variational iteration method. The variational iteration method was developed by Ji-Huan He [1–4]. This method has been used by many researchers to solve problems. The method is applicable for solving a large range of equations, including the acoustic equations. Solutions obtained from this method are approximations of the exact solution.

In this research, we focus on solving the acoustic equations [5–8], in particular the model discussed by LeVeque [5]. To our knowledge, our work is the first in implementing the variational iteration method to solve the acoustic equations. We construct the approximate solutions using the theory of the variational iterations. We then use the Maple software to compute the iterations and generate the plots of the approximate solutions. Several iterations are enough to show that the approximate solutions produced by the variational iteration method are physically correct.

The rest of the paper is organized as follows. First we recall how the variational iteration method works in solving a partial differential equation. Then we construct approximate solutions to the acoustic equations. Afterward, we present some computational results. Finally some concluding remarks are drawn.

2. Variational iteration method

The variational iteration method does not need numerical discretization, as the method is an analytical approach [9, 10]. This method has been proved to be reliable, accurate and effective used to find solutions to differential equations. The method needs an initial value of the differential equations.

Let us illustrate some basic concepts of variational iteration method. Consider the following equation

$$Lu + Nu = g(x). \tag{1}$$

Here L is a linear operator, N is a nonlinear operator, $g(x)$ is a nonhomogeneous term and u is a function of x. From (1), we can construct a correction functional as follows

$$u_{n+1}(x) = u_n(x) + \int_0^x \lambda\{Lu_n(\tau) + N\tilde{u}_n(\tau) - g(\tau)\}d\tau, \quad n \geq 0 \tag{2}$$

with λ being a general Lagrange multiplier. The index n denotes the n-th order iteration, and then $\tilde{u}_n$ is assumed as a restriction variation, that is, $\delta\tilde{u}_n = 0$. We note that integration by parts have the following forms

$$\int \lambda(\tau)u_n'(\tau)d\tau = \lambda(\tau)u_n(\tau) - \int \lambda'(\tau)u_n(\tau)d\tau \tag{3}$$

$$\int \lambda(\tau)u_n''(\tau)d\tau = \lambda(\tau)u_n'(\tau) - \lambda'(\tau)u_n(\tau) + \int \lambda''(\tau)u_n(\tau)d\tau \tag{4}$$

To show how the variational iteration method works, let us consider the following partial differential equation (see Wazwaz [4]):

$$u_x - u_t = 0, \quad u(0,t) = t, \quad u(x,0) = x. \tag{5}$$

We construct the correction functional from (5) to obtain

$$u_{n+1}(x,t) = u_n(x,t) + \int_0^t \lambda(\tau)(u_{n\tau} - \tilde{u}_{nt})d\tau. \tag{6}$$

From (3) and (4) then we obtain the stationary conditions

$$\lambda'(\tau) = 0, \tag{7}$$

$$1 + \lambda(\tau)|_{\tau=t} = 0. \tag{8}$$

The Lagrange multiplier for the above equations is

$$\lambda = -1. \tag{9}$$

Then, we can substitute the Lagrange multiplier (9) into (6) to have the iteration formula

$$u_{n+1}(x,t) = u_n(x,t) - \int_0^t (u_{n\tau} - u_{nt})d\tau, \quad n \geq 0. \tag{10}$$

Iterating (10) we have

$$u_0(x,t) = t, \tag{11}$$

$$u_1(x,t) = x + t, \tag{12}$$

$$u_2(x,t) = x + t, \tag{13}$$

$$u_3(x,t) = x + t, \tag{14}$$

$$\vdots \tag{15}$$

$$u_n(x,t) = x + t. \tag{16}$$

Therefore, the exact solution to (5) is $u(x,t) = x + t$.

3. Acoustic equations

The acoustic equations, in the simplest form, are [11, 12]

$$p_t + u_x = 0, \tag{17}$$

$$u_t + p_x = 0, \tag{18}$$

where $p(x,t)$ is the pressure, $u(x,t)$ is the velocity, t is time variable and x is the space variable. We assume that the initial condition is given by $p(x,0) = 0.1\,\mathrm{sech}^2(0.2x)$ and $u(x,0) = 0$.

We construct a correction functional to solve (17) and (18) using the variational iteration method

$$p_{n+1}(x,t) = p_n(x,t) + \int_0^t \lambda_1(\tau)\left(\frac{\partial p_n(x,\tau)}{\partial \tau} + \frac{\partial \tilde{u}_n(x,\tau)}{\partial x}\right) d\tau, \tag{19}$$

$$u_{n+1}(x,t) = u_n(x,t) + \int_0^t \lambda_2(\tau)\left(\frac{\partial u_n(x,\tau)}{\partial \tau} + \frac{\partial \tilde{p}_n(x,\tau)}{\partial x}\right) d\tau. \tag{20}$$

Here λ_1 and λ_2 are Lagrange multipliers. In (19) we make the functional to be stationary to p_n. Furthermore, in (20) we make the functional to be stationary to u_n, so we have

$$\delta p_{n+1}(x,t) = \delta p_n(x,t) + \delta \int_0^t \lambda_1(\tau)\frac{\partial p_n(x,\tau)}{\partial \tau} d\tau + \delta \int_0^t \lambda_1(\tau)\frac{\partial \tilde{u}_n(x,\tau)}{\partial x} d\tau, \tag{21}$$

$$\delta u_{n+1}(x,t) = \delta u_n(x,t) + \delta \int_0^t \lambda_2(\tau)\frac{\partial u_n(x,\tau)}{\partial \tau} d\tau + \delta \int_0^t \lambda_2(\tau)\frac{\partial \tilde{p}_n(x,\tau)}{\partial x} d\tau, \tag{22}$$

where $\tilde{u}_n$ and $\tilde{p}_n$ are assumed as restriction variations. Noticing that $\delta\tilde{p}_n$ and $\delta\tilde{u}_n = 0$, we obtain

$$\delta p_{n+1}(x,t) = \delta p_n(x,t) + \delta \int_0^t \lambda_1(\tau)\frac{\partial p_n(x,\tau)}{\partial \tau} d\tau, \tag{23}$$

$$\delta u_{n+1}(x,t) = \delta u_n(x,t) + \delta \int_0^t \lambda_2(\tau)\frac{\partial u_n(x,\tau)}{\partial \tau} d\tau. \tag{24}$$

Using integration by parts, (23) and (24) can be written as follows

$$\delta p_{n+1}(x,t) = [1 + \lambda_1(\tau)]\delta p_n(x,t) - \delta \int_0^t \lambda_1'(\tau) p_n(x,t) d\tau, \tag{25}$$

and

$$\delta u_{n+1}(x,t) = [1 + \lambda_2(\tau)]\delta u_n(x,t) - \delta \int_0^t \lambda_2'(\tau) u_n(x,t) d\tau. \tag{26}$$

IOP Conf. Series: Journal of Physics: Conf. Series **856** (2017) 012010 doi:10.1088/1742-6596/856/1/012010

Using (25), we obtain the stationary conditions

$$p_n(\tau,t):\begin{cases}\lambda_1'(\tau)=0,\\ 1+\lambda_1(\tau)=0.\end{cases} \tag{27}$$

Equation (27) leads to

$$\lambda_1(\tau)=-1. \tag{28}$$

Using (26), we obtain the stationary conditions

$$u_n(\tau,t):\begin{cases}\lambda_2'(\tau)=0,\\ 1+\lambda_2(\tau)=0.\end{cases} \tag{29}$$

Equation (29) leads to

$$\lambda_2(\tau)=-1. \tag{30}$$

Now, we obtain the value of the Lagrange multiplier $\lambda_1=\lambda_2=-1$. Substituting that value into (19) and (20), we find

$$p_{n+1}(x,t)=p_n(x,t)-\int_0^t\left(\frac{\partial p_n(x,\tau)}{\partial\tau}+\frac{\partial u_n(x,\tau)}{\partial x}\right)d\tau,\quad n\geq 0, \tag{31}$$

$$u_{n+1}(x,t)=u_n(x,t)-\int_0^t\left(\frac{\partial u_n(x,\tau)}{\partial\tau}+\frac{\partial p_n(x,\tau)}{\partial x}\right)d\tau,\quad n\geq 0. \tag{32}$$

Equations (31) and (32) are the variational iteration formulas for (17) and (18).

4. Computational results

In this section we present some computational experiment to demonstrate the calculations of the variational iteration method. Here we calculate the first, second and third iterations using the variational iteration formulas (31) and (32). All quantities are assumed to have SI units with the MKS system.

For our computational experiment, we are given the initial condition

$$p(x,0)=0.1\,\mathrm{sech}^2(0.2x), \tag{33}$$

and

$$u(x,0)=0. \tag{34}$$

We take hyperbolic secant function for initial pressure and zero function for velocity because these functions are smooth, so they have continuous derivatives. Therefore, they can be used in the demonstration for our method. The amplitude and phase constant in (33) are taken to be 0.1 and 0.2, respectively.

Using the Maple software, we obtain the following results:

$$p_0(x,t)=0.1\,\mathrm{sech}^2(0.2x), \tag{35}$$

$$u_0(x,t)=0, \tag{36}$$

$$p_1(x,t)=0.1\,\mathrm{sech}^2(0.2x), \tag{37}$$

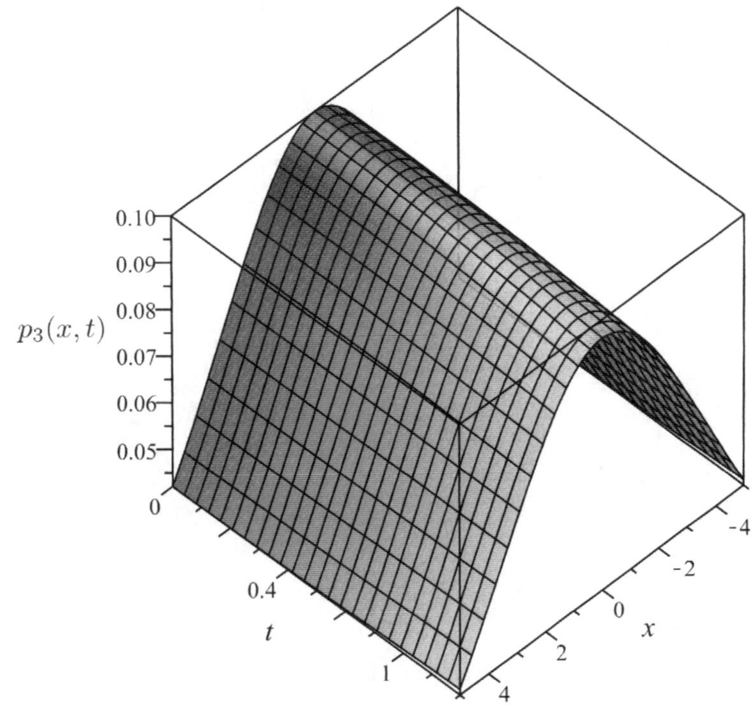

Figure 1. Variational iteration solution for $p_3(x,t)$.

$$u_1(x,t) = 0.04\,\mathrm{sech}(0.2x)^2\tanh(0.2x)t, \tag{38}$$

$$\begin{aligned} p_2(x,t) = {} & 0.1\,\mathrm{sech}(0.2x)^2 + 0.008\,\mathrm{sech}(0.2x)^2\tanh(0.2x)^2\,t^2 \\ & - 0.02\,\mathrm{sech}(0.2x)^2\left(0.2 - 0.2\tanh(0.2x)^2\right)t^2, \end{aligned} \tag{39}$$

$$u_2(x,t) = 0.04\,\mathrm{sech}(0.2x)^2\tanh(0.2x)t, \tag{40}$$

$$\begin{aligned} p_3(x,t) = {} & 0.1\,\mathrm{sech}(0.2x)^2 + 0.008\,\mathrm{sech}(0.2x)^2\tanh(0.2x)^2\,t^2 \\ & - 0.02\,\mathrm{sech}(0.2x)^2\left(0.2 - 0.2\tanh(0.2x)^2\right)t^2, \end{aligned} \tag{41}$$

$$\begin{aligned} u_3(x,t) = {} & 0.04\,\mathrm{sech}(0.2x)^2\tanh(0.2x)t + 0.0011\,\mathrm{sech}(0.2x)^2\tanh(0.2x)^3\,t^3 \\ & - 0.0011\,\mathrm{sech}(0.2x)^2\tanh(0.2x)\left(0.2 - 0.2\tanh(0.2x)^2\right)t^3. \end{aligned} \tag{42}$$

In this paper, we stop our calculations up to the third iteration. The approximate solution for the pressure is $p_3(x,t)$ and the velocity is $u_3(x,t)$. The wave propagation for the pressure is

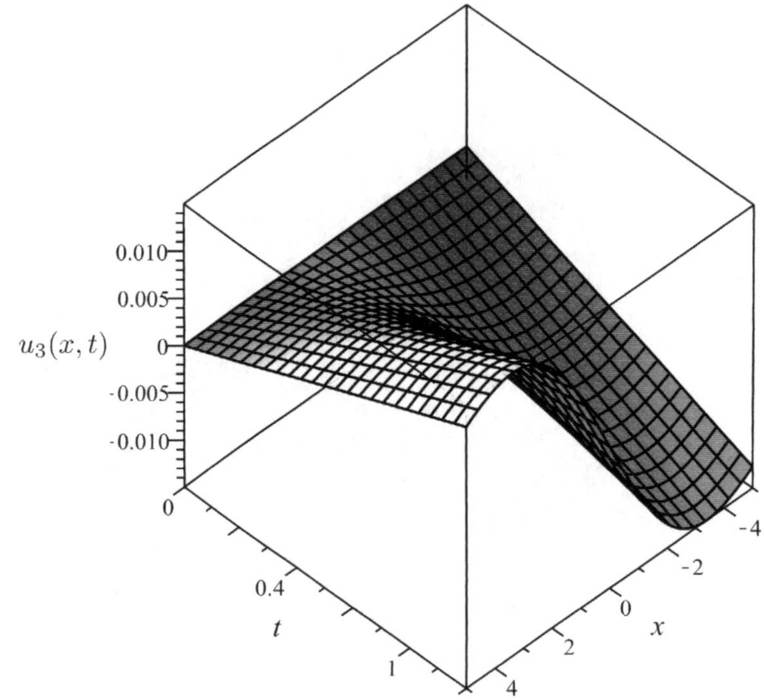

Figure 2. Variational iteration solution for $u_3(x,t)$.

illustrated in figure 1, and for the velocity in figure 2. As time evolves, the pressure from the center of the domain is spread out to the left and to the right. The pressure wave propagating to the left has negative velocity, whereas the one to the right has positive velocity. This behavior is obviously correct physically (moving to the right means that the velocity is positive, and moving to the left means the velocity is negative). In figure 2, the velocity tends to zero for large x and t.

5. Conclusion

We have solved the acoustic equations using the variational iteration method. This method is meshless, which means that it does not need any numerical discretization of the domain. To obtain more accurate results, we just need to do more iterations. Using the variational iteration method, we can compute the solution to the acoustic equations at every value of time and every position of space.

Acknowledgments

The work in this paper was financially supported by Sanata Dharma University, which is gratefully acknowledged by the authors.

References

[1] He J H 1997 *Commun. Nonlinear Sci. Numer. Simul.* **2** 230
[2] He J H 1999 *Int. J. Non Linear Mech.* **34** 699
[3] He J H 2007 *J. Comput. Appl. Math.* **207** 3
[4] Wazwaz A M 2009 *Partial Differential Equations and Solitary Waves Theory* (Beijing: Higher Education Press and Berlin: Springer)

[5] LeVeque R J 2002 *Int. J. Numer. Methods Fluids* **40** 93
[6] LeVeque R J 1997 *J. Comput. Phys.* **131** 237
[7] Fogarty T and LeVeque R J 1998 *Proc. 4th Int. Conf. on Mathematical and Numerical Aspects of Wave Propagation* ed DeSanto J A (Golden) p 603
[8] Fogarty T and LeVeque R J 1999 *J. Acoust. Soc. Am.* **106** 17
[9] Setianingrum P S and Mungkasi S 2016 *AIP Conf. Proc.* **1746** 020057
[10] Mungkasi S and Wiryanto L H 2016 *AIP Conf. Proc.* **1707** 050010
[11] Supriyadi B and Mungkasi S 2016 *Int. J. Multiscale Comput. Eng.* **14** 479
[12] Mungkasi S and Ningrum G I J 2016 *AIP Conf. Proc.* **1746** 020056

IOP Conf. Series: Journal of Physics: Conf. Series **856** (2017) 012011 doi:10.1088/1742-6596/856/1/012011

Role of spin-3/2 hyperon resonances in the kaon photoproduction $\gamma p \to K^+\Lambda$

W D Suciawo, S Clymton and T Mart

Departemen Fisika, FMIPA, Universitas Indonesia, Depok 16424, Indonesia

E-mail: `samsonclymton@gmail.com`

Abstract. We have studied the role of spin-3/2 hyperon resonances in the kaon photoproduction $\gamma p \to K^+\Lambda$ by using an isobar model within the effective Lagrangian approach. This study provides an extension to our previous study, which investigated the effect of nucleon resonances with spins up to 5/2. In the current investigation we have added 7 hyperon resonances with spin-3/2 to the model. The result indicates that the inclusion of spin-3/2 hyperon resonances have a significant effect on this model, i.e., the value of χ^2 can be substantially reduced. Significant improvement can be observed in the polarization observables, especially in the backward angles.

1. Introduction

It has been a long time since the $\gamma p \to K^+\Lambda$ reaction was used as a tool to investigate the properties of kaon and strangeness degrees of freedom. A number of models have been used in these studies to theoretically explain the process of kaon photoproduction. A notorious problem of the spin-coupling inconsistency showed up, especially in the case of higher spin. It is the purpose of our study to alleviate this problem by using the Pascalutsa [1–3] prescription. The prescription has been used in the previous study [4] to investigate the effect of spin-3/2 and spin-5/2 resonance models in the $\gamma p \to K^+\Lambda$ reaction. Note that the previous studies only focused on the nucleon resonances because those are directly associated with their total energy. In the present study we will explore the role of spin-3/2 hyperon resonances in kaon photoproduction by using an isobar model. Since the values of coupling constants are mostly unknown, in our calculation we have to extract them from experimental data. Our present database consists of the cross section and polarization observables data, in which more than 7,400 data points are available from different measurements.

2. The isobar model

The kaon photoproduction reaction investigated in the present study can be formally written as

$$\gamma(k) + p(p) \to K^+(q) + \Lambda(p_\Lambda) \ . \tag{1}$$

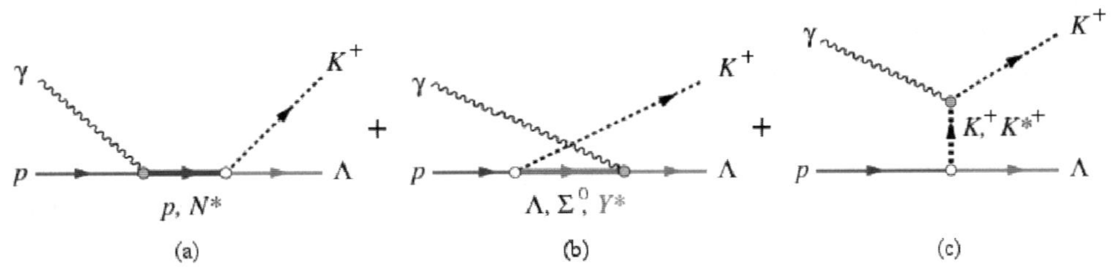

Figure 1. Feynmann diagrams for kaon photoproduction $\gamma + p \to K^+ + \Lambda$. Each diagram describes the (a) s-channel with proton and nucleon resonance intermediate states, (b) u-channel with Λ, Σ^0 and hyperon resonance intermediate states and (c) t-channel with kaon and kaon resonance intermediate states [5].

The process can be described by means of the s-, u- and t-channel Feynman diagrams depicted in figure 1. The Mandelstam variables are defined as

$$s = (p+k)^2 = W^2 \,, \tag{2}$$
$$t = (k-q)^2 \,, \tag{3}$$
$$u = (k-p_\Lambda)^2 \,. \tag{4}$$

In this study, we derive the scattering amplitude by using the Feynmann diagrams given in figure 1. The relevant channel for calculating the scattering amplitude of hyperon resonance is given in figure 1 (b). The obtained scattering amplitude can be decomposed into (see [4], for instance)

$$\mathcal{M}_{fi} = \bar{u}_\Lambda \sum_{i=1}^{6} A_i(s,t,u,k^2)\, M_i\, u_p \,, \tag{5}$$

where $\bar{u}_\Lambda$ and u_p are Λ and proton spinors, respectively. The gauge and Lorentz invariant matrices M_i are given by [6, 7]

$$M_1 = \tfrac{1}{2}\gamma_5(\not{\epsilon}\not{k} - \not{k}\not{\epsilon}) \,, \tag{6}$$
$$M_2 = \gamma_5\left[(2q-k)\cdot\epsilon\; P\cdot k - (2q\cdot k)\cdot k\; P\cdot\epsilon\right] \,, \tag{7}$$
$$M_3 = \gamma_5(q\cdot k\not{\epsilon} - q\cdot\epsilon\not{k}) \,, \tag{8}$$
$$M_4 = i\epsilon_{\mu\nu\rho\sigma}\gamma^\mu q^\nu \epsilon^\rho k^\sigma \,, \tag{9}$$
$$M_5 = \gamma_5(q\cdot\epsilon\; k^2 - q\cdot k\; k\cdot\epsilon) \,, \tag{10}$$
$$M_6 = \gamma_5(k\cdot\epsilon\;\not{k} - k^2\not{\epsilon}) \,, \tag{11}$$

with $P = \frac{1}{2}(p + p_\Lambda)$ and $\epsilon_{\mu\nu\rho\sigma}$ being the four-dimensional Levi-Civita tensor, which is related to the Dirac matrices via $i\epsilon_{\mu\nu\rho\sigma}\gamma^\mu = \gamma_5(-\gamma_\nu\gamma_\rho\gamma_\sigma + g_{\nu\rho}\gamma_\sigma + g_{\rho\sigma}\gamma_\nu - g_{\nu\sigma}\gamma_\rho)$. In the case of photoproduction, we use real photon so that $k\cdot\epsilon = 0 = k^2$. Therefore, both M_5 and M_6 vanish in the calculation. In the case of electroproduction, the virtual photon has $k^2 \neq 0$. As a consequence, both M_5 and M_6 may contribute to the scattering amplitude. For the sake of completeness and future studies, in the present study we derive $A_1, \cdots, A_6$. The cross section and other polarization observables can be calculated by means of the functions $A_1, \cdots, A_6$.

To account for the hadronic structure we include hadronic form factors in all hadronic vertices, where we adopt the method developed by Haberzettl [8] to restore the gauge invariance of amplitudes, which is broken after the inclusion of the form factors. In the present study the hadronic form factor reads

$$F(\Lambda, x) = \frac{\Lambda^4}{\Lambda^4 + (x - m^2)^2} , \tag{12}$$

where Λ is the form factor cutoff, m is the mass of intermediate particle and x refers to one of the Mandelstam variables. We note that this method introduces two more free parameters, which come from the gauge amplitude A_2. Together with the individual form factors, we parameterize this gauge form factor as [8]

$$\tilde{F} = F(\Lambda, s) \sin^2\theta_{\rm had} \cos^2\phi_{\rm had} + F(\Lambda, u) \sin^2\theta_{\rm had} \sin^2\phi_{\rm had} + F(\Lambda, t) \cos^2\phi_{\rm had} , \tag{13}$$

where the values of $\theta_{\rm had}$ and $\phi_{\rm had}$ are obtained from the fit.

Based on the work of Pascalutsa [2], the propagator for spin-3/2 particle can be written as

$$P^{3/2}_{\mu\nu} = \frac{(\not p_\Lambda - \not k) + m_{Y^*}}{3(u - m^2_{Y^*} + i\Gamma_{Y^*} m_{Y^*})} \left[\frac{u}{m^2_{Y^*}} (3P_{\mu\nu} + \gamma^\rho\gamma^\sigma P_{\mu\rho}P_{\nu\sigma}) \right] , \tag{14}$$

where $P_{\mu\nu} = -g_{\mu\nu} + \frac{1}{u}(k - p_\Lambda)_\mu (k - p_\Lambda)_\nu$. We include factor of $u/m^2_{Y^*}$ in (14) in order to regularize the propagator [10,19]. The hadronic vertex is obtained from [1] and [4], whereas the electromagnetic vertex is taken from our previous works [3,4].

Therefore the scattering amplitude can be written as

$$\begin{aligned} \mathcal{M}^{\pm}_{fi} &= \bar{u}_\Lambda \gamma_5 \Big[G^1 \not p_\Lambda (\epsilon^\mu \not k - k^\mu \not\epsilon) + G^2 (k^\mu p_\Lambda \cdot \epsilon - \epsilon^\mu p_\Lambda \cdot k) + G^3 p^\mu_\Lambda (\not\epsilon \not k - \not k \not\epsilon) \Big] \\ &\quad \times (\not p_\Lambda - \not k \pm m_{Y^*})(3P_{\mu\nu} + \gamma^\rho\gamma^\sigma P_{\mu\rho}P_{\nu\sigma}) \Big[p^\nu (\not k - \not p_\Lambda) \Big] u_p , \end{aligned} \tag{15}$$

where

$$G^i = \frac{u\, g_{KY^*p}\, g^i}{3m^6_{Y^*}(u - m^2_{Y^*} + i\Gamma_{Y^*} m_{Y^*})} , \tag{16}$$

and

$$g^1 = -2ig^a_{\gamma\Lambda Y^*} + ig^c_{\gamma\Lambda Y^*} - g^d_{\gamma\Lambda Y^*} , \tag{17}$$
$$g^2 = -2ig^a_{\gamma\Lambda Y^*} - g^b_{\gamma\Lambda Y^*} + 2ig^c_{\gamma\Lambda Y^*} - 2g^d_{\gamma\Lambda Y^*} , \tag{18}$$
$$g^3 = ig^a_{\gamma\Lambda Y^*} - ig^c_{\gamma\Lambda Y^*} . \tag{19}$$

By decomposing (15) into M_i matrices, we obtain the amplitude A_i

$$\begin{aligned} A_1 &= \Big[\tfrac{1}{2} m_\Lambda (m_p + m_\Lambda) \{ 4c_p + 3u \pm 3m_\Lambda m_{Y^*} \mp m_p m_{Y^*} + m_\Lambda (\pm \tfrac{2}{u} c_p m_{Y^*} + m_p) \} \mp 3b_p m_\Lambda m_{Y^*} \\ &\quad - m_\Lambda (b_\Lambda - k^2)(\pm \tfrac{2}{u} c_p m_{Y^*} + m_p) - 2m_\Lambda m_p (u \mp m_p m_{Y^*}) - 2m_\Lambda c_p (m_p \mp m_{Y^*}) \Big] G^1 \\ &\quad + \Big[\tfrac{1}{2}(m_p + m_\Lambda) \{ \pm 3m_{Y^*}(b_p - b_\Lambda) - k^2 (\pm \tfrac{2}{u} c_p m_{Y^*} + m_p) + m_p (u \mp m_\Lambda m_{Y^*}) \\ &\quad + c_p (m_\Lambda \mp m_{Y^*}) \} + b_\Lambda (\pm m_p m_{Y^*} - c_p) \Big] G^2 + \Big[6(b_p - c_p)(u \pm m_p m_{Y^*}) - 2c_\Lambda (4c_p \\ &\quad \pm \tfrac{2}{u} c_p m_p m_{Y^*} \mp m_p m_{Y^*} + m_p^2) - 4b_\Lambda (\pm m_p m_{Y^*} - c_p) + 2m_\Lambda \{ m_p (u \mp m_p m_{Y^*}) \\ &\quad + c_p (m_p \mp m_{Y^*}) \} - 2(m_p + m_\Lambda) \{ m_p (u \mp m_\Lambda m_{Y^*}) + c_p (m_\Lambda \mp m_{Y^*}) \} \Big] G^3 , \end{aligned} \tag{20}$$

$$\begin{aligned} A_2 &= \frac{1}{t - m_K^2} \Big[\{ \mp 6m_\Lambda m_{Y^*}(b_\Lambda - b_p) - m_\Lambda k^2 (m_p \pm \tfrac{2}{u} c_p m_{Y^*} \mp 3m_{Y^*}) \} G^1 + \{ k^2 \{ m_\Lambda (m_p \\ &\quad \pm \tfrac{2}{u} c_p m_{Y^*}) - 3c_p \} - 3(b_\Lambda - b_p)(u \mp m_\Lambda m_{Y^*}) \} G^2 + \{ 4k^2 (\pm m_p m_{Y^*} - c_p) \} G^3 \Big] , \end{aligned} \tag{21}$$

IOP Conf. Series: Journal of Physics: Conf. Series **856** (2017) 012011 doi:10.1088/1742-6596/856/1/012011

$$
\begin{aligned}
A_3 &= \Big[\tfrac{1}{2}m_\Lambda\{3u \pm 3m_\Lambda m_{Y^*} \mp 3m_p m_{Y^*} - m_\Lambda(\pm\tfrac{2}{u}c_p m_{Y^*} + m_p)\}\Big]G^1 + \Big[\tfrac{1}{2}\{\mp 3m_{Y^*}(b_p \\
&\quad + b_\Lambda) + k^2(\pm\tfrac{2}{u}c_p m_{Y^*} + m_p) - m_p(u \mp m_\Lambda m_{Y^*}) - c_p(m_\Lambda \mp m_{Y^*})\}\Big]G^2 \\
&\quad + \Big[\mp 6m_{Y^*}(b_p - c_p) - 2(\pm c_p m_{Y^*} - u m_p) + 2c_\Lambda(\pm\tfrac{2}{u}c_p m_{Y^*} + m_p)\Big]G^3 , \qquad (22) \\
A_4 &= \Big[\tfrac{1}{2}m_\Lambda\{3u \pm 3m_\Lambda m_{Y^*} \pm 3m_p m_{Y^*} + m_\Lambda(\pm\tfrac{2}{u}c_p m_{Y^*} + m_p)\}\Big]G^1 + \Big[\tfrac{1}{2}\{\pm 3m_{Y^*}(b_p \\
&\quad - b_\Lambda) - k^2(\pm\tfrac{2}{u}c_p m_{Y^*} + m_p) + m_p(u \mp m_\Lambda m_{Y^*}) + c_p(m_\Lambda \mp m_{Y^*})\}\Big]G^2 \\
&\quad + \Big[\pm 6m_{Y^*}(b_p - c_p) + 2(\pm c_p m_{Y^*} - u m_p) - 2c_\Lambda(\pm\tfrac{2}{u}c_p m_{Y^*} + m_p)\Big]G^3 , \qquad (23) \\
A_5 &= \frac{1}{t - m_K^2}\Big[\{\mp 3m_\Lambda m_{Y^*}(\tfrac{1}{2}k^2 + b_p - b_\Lambda) - m_\Lambda(\pm\tfrac{2}{u}c_p m_{Y^*} + m_p)(\tfrac{3}{2}k^2 - 2b_\Lambda)\}G^1 \\
&\quad + \{\tfrac{3}{2}(b_p + b_\Lambda)(u \mp m_\Lambda m_{Y^*}) - (2b_\Lambda + \tfrac{1}{2}k^2)\{m_\Lambda(\pm\tfrac{2}{u}c_p m_{Y^*} + m_p) - 3c_p\}\}G^2 \\
&\quad - \{4(2b_\Lambda + \tfrac{1}{2}k^2)(\pm m_p m_{Y^*} - c_p)\}G^3\Big] , \qquad (24) \\
A_6 &= \Big[\tfrac{1}{2}m_\Lambda\{3(u \pm m_\Lambda m_{Y^*}) + 4c_p \mp m_p m_{Y^*} + m_\Lambda(\pm\tfrac{2}{u}c_p m_{Y^*} + m_p)\}\Big]G^1 + \Big[\mp\tfrac{3}{2}m_{Y^*} \\
&\quad \times(b_p + b_\Lambda) - (b_\Lambda - \tfrac{1}{2}k^2)(\pm\tfrac{2}{u}c_p m_{Y^*} + m_p) - \tfrac{1}{2}m_p(u \mp m_\Lambda m_{Y^*}) \\
&\quad - \tfrac{1}{2}c_p(m_\Lambda \mp m_{Y^*})\Big]G^2 + \Big[2\{m_p(u \mp m_\Lambda m_{Y^*}) + c_p(m_\Lambda \mp m_{Y^*})\}\Big]G^3 , \qquad (25)
\end{aligned}
$$

using some notation

$$
\begin{aligned}
b_p &= p \cdot k = \tfrac{1}{2}(s - k^2 - m_p^2) , \\
b_\Lambda &= p_\Lambda \cdot k = \tfrac{1}{2}(k^2 + m_\Lambda^2 - u) , \\
c_p &= (k - p_\Lambda) \cdot p , \\
c_\Lambda &= (k - p_\Lambda) \cdot p_\Lambda .
\end{aligned}
$$

to simplify the result of calculation.

The background terms consist of the standard s-,u- and t-channel Born terms including two vector mesons $K^{*+}(892)$ and $K_1(1270)$ [4, 5, 9, 11–17]. Two spin-1/2 hyperon resonances are also included in the model because they have been shown to be important to maintain a reasonable value of the hadronic form-factor cutoff of the Born terms [5]. Meanwhile, all nucleon resonances with spins up to 5/2 listed in the Review of Particle Properties of Particle Data Group (PDG) [18], with at least ** rating, are also included as in our previous study [4]. In the present study, we include all spin-3/2 of Λ and Σ resonances listed by the PDG [18] with more than ** rating, as can be seen in table 1. This * rating represents evidence of existence of the particle, where **** means evidence of existence is certain and has been proven and explored by many experiments, while * means evidence of existence is poor. In this study, we consider particle with rating at least **, hence quite fair evidence of existence to be included in our calculation.

3. Results and discussion

The result of fitting the calculated observables to experimental data is summarized in table 2, where we compare the numerical results of the background properties obtained in the present work with those of the previous study [4].

Obviously, the inclusion of the spin-3/2 hyperon resonances significantly decreases the value of χ^2/N :

$$
\frac{\chi^2}{N} = \frac{1}{N_{data} - N_{param}} \sum_{i=1}^{N_{data}} \left[\frac{\sigma_i(exp.) - \sigma_i(calc.)}{\delta\sigma_i}\right]^2 \qquad (26)
$$

IOP Conf. Series: Journal of Physics: Conf. Series **856** (2017) 012011 doi:10.1088/1742-6596/856/1/012011

Table 1. List of the hyperon resonances with spin 3/2 included in this study. This table shows the spin, parity, status, mass and width of the hyperon resonances. Entries are taken from the Review of Particle Properties of PDG [18]. J^P represents spin-parity of the particle, and Status represents evidence of existence of the particle, where more stars mean their existence is more certain.

Particle	J^P	Status	Mass (MeV)	Width (MeV)
$\Lambda(1520)$	$3/2^-$	****	1519.5 ± 1.0	15.6 ± 1.0
$\Lambda(1690)$	$3/2^-$	****	1690 ± 5	60 ± 10
$\Lambda(1890)$	$3/2^+$	****	1890^{+10}_{-40}	100^{+100}_{-40}
$\Sigma(1385)$	$3/2^+$	****	1383.7 ± 1.0	36 ± 5
$\Sigma(1670)$	$3/2^-$	****	1670^{+15}_{-5}	60 ± 20
$\Sigma(1940)$	$3/2^-$	***	1940^{+10}_{-40}	220^{+80}_{-70}
$\Sigma(2080)$	$3/2^+$	**	2080 ± 104	186 ± 48

where N is number of degree of freedom, N_{data} is number of experimental data, N_{param} is number of free parameters that we will fit with experimental data, $\sigma_i(exp.)$ is the i-th observable value from expriments, $\sigma_i(calc.)$ is i-th observable value from our calculation using isobar model, and $\delta\sigma_i$ is the i-th relative error of experimental data. Therefore, the smaller the value of χ^2/N, the better our calculation fitting experimental results.

Figure 2 compares the calculated total cross section obtained in this work and those of the previous studies with various experimental data obtained from different experiments. Clearly, the inclusion of spin-3/2 hyperon resonances in our calculation improves the agreement between the calculated total cross section and the experimental ones, especially around the first peak as well as at higher energies (≥ 2.25 GeV).

Table 2. The background coupling constants and hadronic cutoff obtained in our previous study [4] (without spin-3/2 hyperon resonances) and those obtained in the present study (with spin-3/2 hyperon resonances). Notation of the parameters is as in our previous study [4].

Parameters	without [4]	with
$G_{K\Lambda N}/\sqrt{4\pi}$	−3.00	−3.87
$G_{K\Sigma N}/\sqrt{4\pi}$	1.27	1.30
$G^V_{K^*}/4\pi$	0.15	−0.11
$G^T_{K^*}/4\pi$	0.26	0.26
$G^V_{K_1}/4\pi$	1.46	−0.42
$G^T_{K_1}/4\pi$	0.07	−0.61
$G_{\Lambda(1600)}/4\pi$	8.41	−10.0
$G_{\Lambda(1810)}/4\pi$	−9.61	3.04
Λ_B(GeV)	0.70	0.94
Λ_R(GeV)	1.31	1.35
$\theta_{\rm had}$(deg)	130	90.0
$\phi_{\rm had}$(deg)	177	39.7
χ^2/N	1.58	1.17

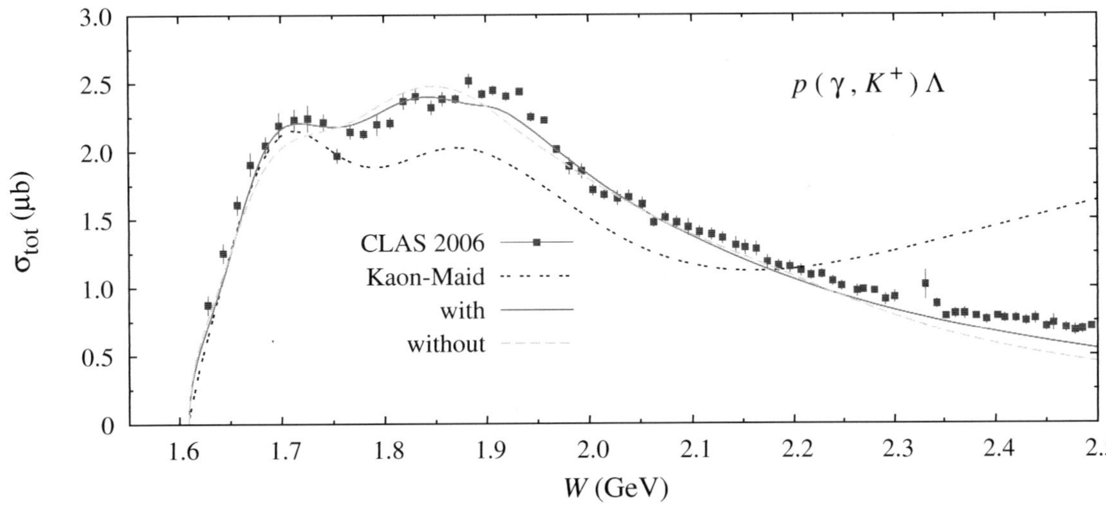

Figure 2. Total cross sections obtained in the present study, which includes the spin-3/2 hyperon resonances (solid red line), compared with the calculation without the spin-3/2 hyperon resonances (previous study, dashed green line) [4] and the prediction of Kaon-Maid (dotted black line) [9]. Experimental data (solid blue square) are taken from the CLAS Collaboration [20].

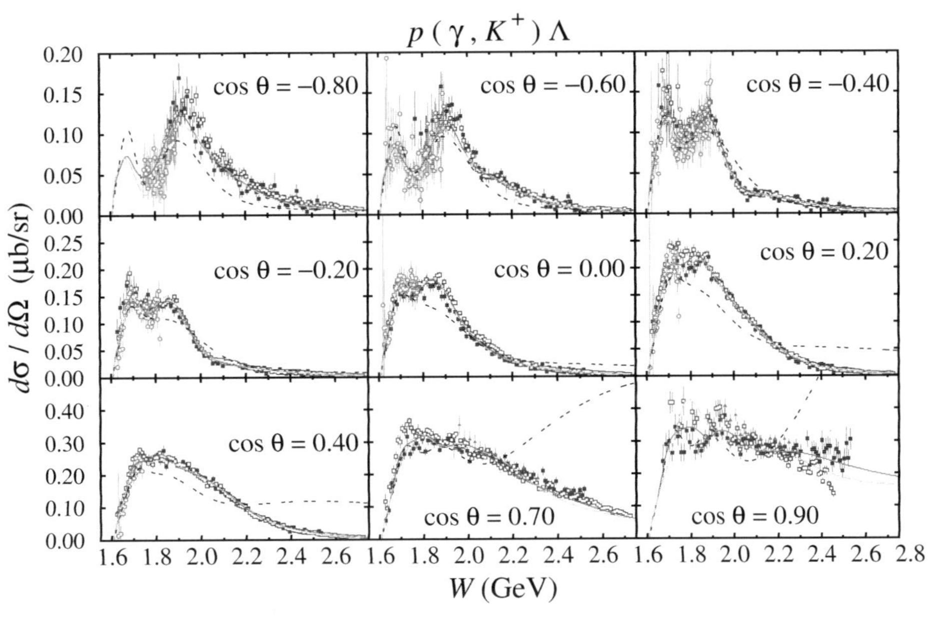

Figure 3. Energy distributions of differential cross section for different values of kaon scattering angle. Experimental data are obtained from the CLAS Collaboration (solid squares and open squares) [20, 21], Crystal Ball Collaboration (open circles) [23] and LEPS Collaboration (solid triangles) [22]. Notation of the curves is as in figure 2.

IOP Conf. Series: Journal of Physics: Conf. Series **856** (2017) 012011 doi:10.1088/1742-6596/856/1/012011

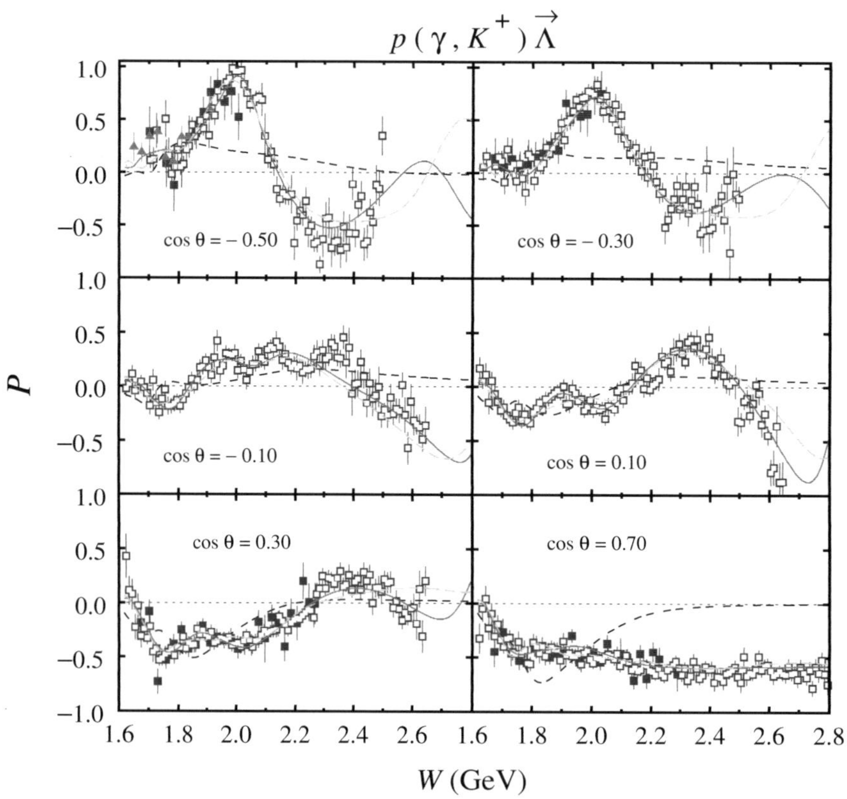

Figure 4. Energy distributions of Λ recoil polarization for different values of kaon scattering angle. Experimental data are from the CLAS Collaboration (solid squares and open squares) [20, 21] and the GRAAL Collaboration (solid triangle) [24]. Notation of the curves is as in figure 3.

Figure 3 shows the energy distributions of differential cross section for different values of kaon scattering angles, where we compare experimental data with the calculated differential cross sections obtained from our previous study [4], Kaon-Maid [9] and present investigation. From figure 3 we can conclude that the inclusion of spin-3/2 hyperon resonances has a strong impact at the backward and forward angles. This is understandable since the corresponding Mandelstam variable in the hyperon propagator is u, which depends on $\cos\theta$. At $\cos\theta = -0.80$ the inclusion of hyperon resonances can substantially improve the agreement with experimental data around the second peak of the cross section. A similar phenomenon also happens at $\cos\theta = 0.90$, where the improvement seems to be also significant.

Polarization observable has been long known as a decisive constraint for phenomenological models. In the case of kaon photoproduction, the well known and very often used polarization observable is the Λ recoil polarization. The result of our calculation together with those of previous studies are shown in figure 4. In this figure, we can also see that the inclusion of spin-3/2 hyperon resonances provides the best strategy to improve our previous study. Especially at higher energies, the agreement with experimental data is significantly improved. Even though the experimental data is quite scattered in higher energy, this result is able to determine those data better than our previous study.

4. Summary and conclusion

We have studied kaon photoproduction by using an isobar model with spin-3/2 hyperon resonances included. This inclusion has been shown to decrease the value of χ^2 significantly. By looking at the polarization observable, we conclude that the spin-3/2 hyperon resonances play an important role especially at the backward angles of kaon photoproduction.

Acknowledgments

This work has been partly supported by the Research-Cluster-Grant-Program of the University of Indonesia, under contract No. 1862/UN.R12/HKP.05.00/2015.

References

[1] Pascalutsa V 1998 *Phys. Rev.* D **58** 096002
[2] Pascalutsa V 2001 *Phys. Lett.* B **503** 85
[3] Pascalutsa V and Timmermans R 1999 *Phys. Rev.* C **60** 042201
[4] Mart T, Clymton S and Arifi A J 2015 *Phys. Rev.* D **92** 094019
[5] Mart T and Nurhadiansyah N 2013 *Few-Body Syst.* **54** 1729
[6] Deo B B and Bisoi A K 1974 *Phys. Rev.* D **9** 288
[7] Dennery P 1961 *Phys. Rev.* **124** 2000
[8] Haberzettl H, Bennhold C, Mart T and Feuster T 1998 *Phys. Rev.* C **58** R40
[9] Mart T and Bennhold C 1999 *Phys. Rev.* C **61** 012201
[10] Vrancx T, De Cruz L, Ryckebusch J and Vancraeyveld P 2011 *Phys. Rev.* C **84** 045201
[11] Mart T and Sulaksono A 2006 *Phys. Rev.* C **74** 055203
[12] Mart T 2010 *Phys. Rev.* C **82** 025209
[13] Mart T and Kholili M J 2012 *Phys. Rev.* C **86** 022201
[14] Mart T and Sari A K 2013 *Mod. Phys. Lett.* A **28** 1350054
[15] Mart T 2013 *Phys. Rev.* C **87** 042201
[16] Mart T 2014 *Phys. Rev.* C **90** 065202
[17] Syukurilla L and Mart T 2015 *Int. J. Mod. Phys.* E **24** 1550008
[18] Olive K A *et al* 2014 *Chin. Phys.* C **38** 090001
[19] Shklyar V, Lenske H and Mosel U 2010 *Phys. Rev.* C **82** 015203
[20] Bradford R *et al* 2006 *Phys. Rev.* C **73** 035202
[21] McCracken M E *et al* 2010 *Phys. Rev.* C **81** 025201
[22] Sumihama M *et al* 2006 *Phys. Rev.* C **73** 035214
[23] Jude T C *et al* 2014 *Phys. Lett.* B **735** 112
[24] Lleres A *et al* 2007 *Eur. Phys. J.* A **31** 79
[25] Lleres A *et al* 2009 *Eur. Phys. J.* A **39** 149
[26] Bradford R *et al* 2007 *Phys. Rev.* C **75** 035205

IOP Conf. Series: Journal of Physics: Conf. Series **856** (2017) 012012 doi:10.1088/1742-6596/856/1/012012

Finding a Hadamard matrix by simulated annealing of spin vectors

Andriyan Bayu Suksmono

School of Electrical Engineering and Informatics, Institut Teknologi Bandung, Jl. Ganesha 10, Bandung, Indonesia

E-mail: suksmono@stei.itb.ac.id, suksmono@yahoo.com

Abstract. Reformulation of a combinatorial problem into optimization of a statistical-mechanics system enables finding a better solution using heuristics derived from a physical process, such as by the simulated annealing (SA). In this paper, we present a Hadamard matrix (H-matrix) searching method based on the SA on an Ising model. By equivalence, an H-matrix can be converted into a seminormalized Hadamard (SH) matrix, whose first column is unit vector and the rest ones are vectors with equal number of -1 and $+1$ called SH-vectors. We define SH spin vectors as representation of the SH vectors, which play a similar role as the spins on Ising model. The topology of the lattice is generalized into a graph, whose edges represent orthogonality relationship among the SH spin vectors. Starting from a randomly generated quasi H-matrix Q, which is a matrix similar to the SH-matrix without imposing orthogonality, we perform the SA. The transitions of Q are conducted by random exchange of $\{+,-\}$ spin-pair within the SH-spin vectors that follow the Metropolis update rule. Upon transition toward zeroth energy, the Q-matrix is evolved following a Markov chain toward an orthogonal matrix, at which the H-matrix is said to be found. We demonstrate the capability of the proposed method to find some low-order H-matrices, including the ones that cannot trivially be constructed by the Sylvester method.

1. Introduction

A Hadamard matrix (H-matrix) is an orthogonal binary matrix whose entries are -1 or $+1$. The matrix had been studied by Sylvester [1] and was later realized by Jacques Hadamard during his investigation on maximal determinant problem [2]. Unique properties of the H-matrix attracts many mathematicians, scientists and engineers to study it more intensively. The orthogonality and binariness of the matrix enable various kinds of practical applications. In digital communication area, the Hadamard-Walsh code has been used in Code Division Multiple Access (CDMA) systems, such as the IS-95 or CDMA2000 [3, 4]. H-matrix has also been used to construct an error-correction code, which is capable of correcting large errors suffered by the communicated data. Due to this capability, in 1971 NASA employed the Hadamard code for the Mariner 9 space probe to send digital images of Planet Mars across the deep space toward a receiver on Earth.

One of the most important problems in the study of the H-matrix is its existence. It is well known that 2^k-order H-matrices for any positive integer k exist, in which the Sylvester method can be used for the construction. Various H-matrix construction methods for order other than 2^k, such as Paley's finite field [5], Dade-Goldberg's group permutation [6], Williamson method [7],

IOP Conf. Series: Journal of Physics: Conf. Series **856** (2017) 012012 doi:10.1088/1742-6596/856/1/012012

Bush's finite projective plane [8,9] and Wallis's orthogonal design [10], have also been proposed. However, there is no general method for constructing the $4k$-th order H-matrix. The Hadamard matrix conjecture states that there is a $4k$-order H-matrix for every positive integer k. At present, the largest-known H-matrix for a general $4k$-order is 428, which was discovered by Karaghani and Tayfeh-Rezaie [11].

In principle, provided that the Hadamard conjecture is true, one can find a $4k$-order H-matrix by performing orthogonality tests to all of the $4k$-order binary matrices. The total number N_B of such $4k$-order binary matrix is

$$N_B\left(4k\right) = 2^{16k^2}. \tag{1}$$

Therefore, in the worst-case condition, the search space is of exponential order. On the other hand, verifying that a binary $4k$-order matrix is Hadamard (or not), one needs only a matrix multiplication whose complexity is of polynomial order. Consequently, finding a Hadamard matrix is a nondeterministic polynomial (NP) problem.

The usage of SA to solve hard optimization problems was initiated by Kirkpatrick *et al* [12] and Černý [13], who realized the similarity and a deep connection between combinatorial problems and the statistical mechanics. The optimization method proposed by these authors employs Monte Carlo algorithm with Metropolis update rule [14]. For other practical applications, the SA has also been used in image restorations: restoring the real-valued images [15] and the complex-valued ones [16].

In this paper, we present a method of finding a Hadamard matrix by using SA. The distinctive feature in our method is, instead of using binary values on an Ising lattice, we consider a special binary vectors or spin vectors, which have balanced numbers of -1 and $+1$ spins. By starting from a system of a set of randomly generated $(4k-1)$ spin vectors, the SA seeks for a configuration with global minimum energy of the spin-vector system. When the ground state is achieved, the set of vectors becomes orthogonal and a $4k$-order H-matrix can be constructed. At this condition, we say that the H-matrix has been found.

The rest of the paper is organized as follows. In Section 2, we briefly discuss the basic properties of an H-matrix, SH vectors and SH-spin vectors employed throughout the paper. The proposed method is formulated in Section 3, where an SA algorithm for the SH spin-vectors case is sketched. Search examples of a few low-order H-matrices, including the ones that cannot be constructed by the Sylvester method, are presented in Section 4. The paper will be concluded in Section 5.

2. The Hadamard matrix and seminormalized Hadamard spin vectors

2.1. SH matrix, SH vectors and quasi H-matrix

An m-order H-matrix is an $m \times m$ orthogonal matrix whose entries are either -1 or 1. We will be writing explicitly the signs of the entries so that the number 1 is written as $+1$ for clarity of explanations. We also assume that $m = 4k$, with k a positive integer, throughout the paper. The orthogonality of the H-matrix means that the following relationship is held

$$H^T H = mI \tag{2}$$

where I is an $m \times m$ identity matrix. For an H-matrix, reordering (exchanging or permuting) the rows (or columns), transposition and negation (multiplication by -1) of the rows (or columns) yield another H-matrix, i.e., these operations do not change the status of the H-matrix. The H-matrices obtained by such operations are called equivalent H-matrices, and the operations are called equivalent operations.

By the equivalent operations, an H-matrix can be normalized or seminormalized. An H-matrix is normalized, shortly called an NH-matrix, if the entries of both the first column and the first row are +1. Whereas an H-matrix is seminormalized, written as an SH-matrix, if the entries of either its first column or first rows, are +1. For consistency, we will choose the SH-matrix as an H-matrix whose first column entries are +1 or a unit vector. Furthermore, we will write +1 entries as + and the −1 as − for conciseness. The following is equivalent operations to an H-matrix that yield an SH-matrix and an NH-matrix.

$$H_1 = \begin{pmatrix} + & + & - & - \\ + & - & - & + \\ - & - & - & - \\ + & - & + & - \end{pmatrix} \rightarrow H_2 = \begin{pmatrix} + & + & - & - \\ + & - & - & + \\ + & + & + & + \\ + & - & + & - \end{pmatrix} \rightarrow H_3 = \begin{pmatrix} + & + & + & + \\ + & - & - & + \\ + & + & - & - \\ + & - & + & - \end{pmatrix}$$

In this example, H_1, H_2 and H_3 are equivalent H-matrices. In particular, H_2 is an SH-matrix that is obtained by negating the third row of H_1, whereas H_3 is an NH-matrix that is obtained by exchanging the first row with the third one of H_2.

For an SH-matrix, since the first column is a unit vector, the orthogonality condition in (2) implies that the other column vectors must have a balanced number of −1 and +1, i.e., they are $4k$-order vectors whose $2k$ entries are +1 and the rest of $2k$ number of entries are −1. Based on basic counting, the number of $4k$-order SH vector N_V is equal to the number of ways of arranging $2k$ number of objects, i.e., −1 into $4k$ positions. Then, we obtain the following result

$$N_V\,(4k) = C(4k, 2k) = \frac{4k!}{(2k)!(2k)!}. \tag{3}$$

In the previous example, the 2nd, 3rd and 4th columns of H_2 and H_3 are SH vectors. The complete list of 4-order SH vectors, which according to (3) will consist of $N_V = 6$ distinct vectors, is as follows:

$\vec{v}_1 = (\,+ \;\; + \;\; - \;\; -\,)^T,\ \vec{v}_2 = (\,- \;\; - \;\; + \;\; +\,)^T,\ \vec{v}_3 = (\,+ \;\; - \;\; + \;\; -\,)^T,$
$\vec{v}_4 = (\,+ \;\; - \;\; - \;\; +\,)^T,\ \vec{v}_5 = (\,- \;\; + \;\; + \;\; -\,)^T,\ \vec{v}_6 = (\,- \;\; + \;\; - \;\; +\,)^T.$

A pair of (randomly generated) SH-vectors is generally not orthogonal. In analogy to the SH-matrix, we define a square binary matrix Q, whose first column is a unit vector and the rest ones are SH-vectors, that is called a quasi H-matrix or the Q-matrix. In general, Q is nonorthogonal, and at a very special condition, when it is orthogonal, Q becomes an H-matrix. It can be shown that the number of unique $4k$-order quasi H-matrix $N_{QU}(4k)$ can be approximated as follows

$$N_{QU}(4k) \approx \left(\frac{2^{4k}}{8k^{3/2}}\right)^{4k}. \tag{4}$$

Compared with the number of binary matrix of the corresponding order given by (2), the number of unique Q is relatively smaller.

2.2. SH-spin vectors

Usually, an Ising lattice consists of two kinds of spin, i.e., spin-up +1 and spin-down −1. We need to define a more general spin notation that consists of a group of $4k$ number of spins, and treat it as a single entity. The purpose of defining such an object is to represent the SH-vectors as elements (entries) of H-matrix on Ising model. The main idea of finding H-matrix by the SA is evolving a binary nonorthogonal Q-matrix through a Markov chain by randomly changing the vectors from one SH-vector to another SH-vector, followed by detecting the orthogonality of the matrix. When the orthogonality condition is achieved, the system is said to reach a ground state. Since a particular SH vector should only change into another SH-vector, the single-spin flip at a time suggested in [17] is not sufficient; instead, a pair of − and + spins in the SH spin vector should be flipped or exchanged. Therefore, the SH-spin vectors are objects on the Ising lattice that posses the following properties:

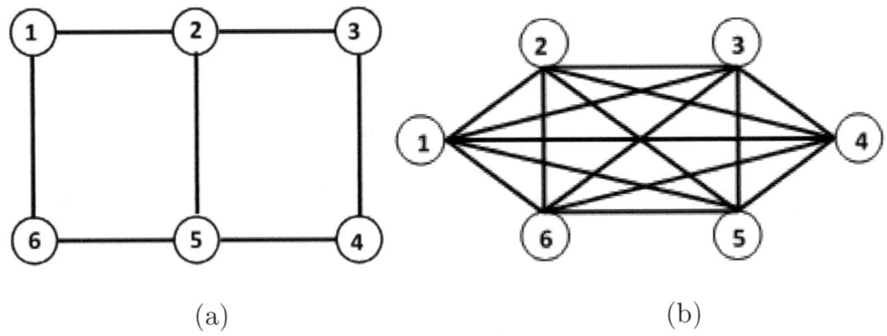

Figure 1. Graphical illustration of the Ising model and connection among the spins: (a) nearest-neighbor connections, and (b) fully connected spins.

- It is a vector that consists of $2k$ number of spin-up ($+1$) and $2k$ number of spin down (-1)
- The change or transition of a spin vector should involve a pair of spin-up and spin-down so that the SH-vector property is preserved.

For illustration, the following transition of exchanging the second spin with the third one $(+\ +\ -\ -)^T \rightarrow (+\ -\ +\ -)^T$ is allowed; however, the following transition $(+\ +\ -\ -)^T \rightarrow (+\ +\ +\ -)^T$ of flipping only the third spin is forbidden since the result is not an SH-vectors, i.e., the numbers of -1 and $+1$ are not balanced anymore.

3. Simulated annealing of SH spin-vectors

3.1. Scalar spin with single-spin flip

Ising model is basically a statistical mechanics model of ferromagnetism, which exhibits not only phase-transition phenomenon (in two-dimensional, Peierls [18]) but also exactly solvable equations for some particular cases [19, 20]. Recently, De las Cuevas and Cubitt proposed the two-dimensional Ising model as a universal model that is capable of capturing all aspects of classical spin physics [21].

In an Ising model, spins are normally arranged on a regular grid (lattice), and each of the spins interacts with its neighbors, as shown in figure 1(a). The energy of a particular spin configuration s is given by

$$E(s) = -\sum_{\langle i,j\rangle} J_{ij} s_i s_j - \mu \sum_j h_j s_j, \tag{5}$$

where $\langle i, j\rangle$ denotes the summation over the products of different s_i and s_j, J_{ij} is the interaction strength or coupling between spins at site i and the one at site j, while h_j is the interaction strength between external magnetic field μ with a spin s_j. When considering nearest-neighbor interaction, the nonzero value of J_{ij} is generally applied only to the directly or neighboring connected pairs so that in figure 1(a), for example, the $J_{1,2}$ and $J_{1,6}$ are nonzero, while $J_{1,3} = 0$. In this paper, we consider a more general arrangement of the spins in terms of configuration and interaction. The spins are represented by vertices of a graph, whereas the interactions are represented by edges of the graph. Fully connected spins will be represented by a complete graph, as shown in figure 1(b).

The configuration of the Ising model is initialized by a randomly generated Q-matrix Q. Since the SA minimization should evolve Q from a high-energy state down to the ground state, where Q becomes orthogonal, the ground state can be considered as a condition of which the system energy becomes zero, whereas nonorthogonality of Q should be associated with a positive

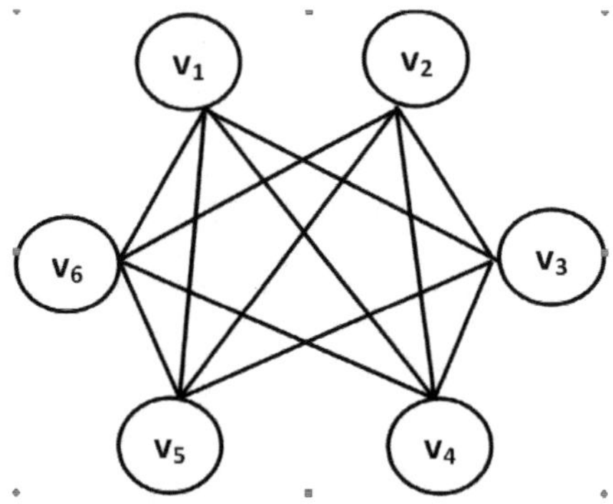

Figure 2. Orthogonality graph of the 4-order SH vectors.

(higher) energy state. Therefore, in a single-flip spin scheme, we may define the energy of an $(m = 4k)$-order Q, which indicates the deviation of Q from an orthogonal matrix, as follows. First, we define an indicator matrix $D(Q) = Q^T Q$ whose entries are denoted by $d_{ij}(Q)$. Then, the energy is given by

$$E(Q) = \sum_i \sum_j |d_{ij}(Q)| - m^2, \tag{6}$$

where $|\cdot|$ denotes the absolute value. If Q is orthogonal, then the first term on the right-hand side expresses the sum of entries of a diagonal matrix; hence $E(Q) = 0$ as desired.

3.2. SH-vector spin with pair-of-spin flip

Alternatively, we can consider an orthogonality graph of $4k$-order SH vectors, such as displayed in figure 2 for a 4-order SH vectors. Each of the $(4k - 1)$ vertices of the complete subgraph can be used to construct an SH-matrix H, such as $\{v_1, v_3, v_4\}$, by inserting a unit vector as the first column, whereas this set of SH vectors belongs to other columns. In the SA, we use a set of randomly generated $(4k - 1)$ SH-vectors, which are then connected as a $(4k - 1)$ complete graph, to construct an SH-Ising vectors lattice. This is followed by evolving them by using SA algorithm so that they subsequently become orthogonal to each other. The energy of such a configuration can be written as:

$$E(\vec{v}) = \sum_{i \neq j} |\vec{v}_i \cdot \vec{v}_j| + \sum_i \left| \vec{1} \cdot \vec{v}_i \right| . \tag{7}$$

The second term in (7) ensures that the evolution maintains the vectors v_i of being an SH-vectors. Since we restrict v_i to only SH-vectors, the spin flip on the SH-spin vector should not change the numbers of $+1$ and -1 entries. It can be achieved only if we flip them in pairs, or exchange the position of the pair.

Figure 3 displays the connections among SH-spin vectors. In figure 3(a), the connections are drawn among the individual spins, whereas in figure 3(b) they are represented by connection among SH-spin vectors. The pseudocode listed in Algorithm 1 illustrates the proposed SH-spin vectors SA method.

In the SA algorithm, we have used a scheduled annealing. Initially, the system is set to a total randomness by setting the value of transition threshold of the Metropolis to 0.5. Then, the threshold is increased subsequently such that it reaches 1.0 at the end of the iteration. A detailed example of the schedule is described in the following section.

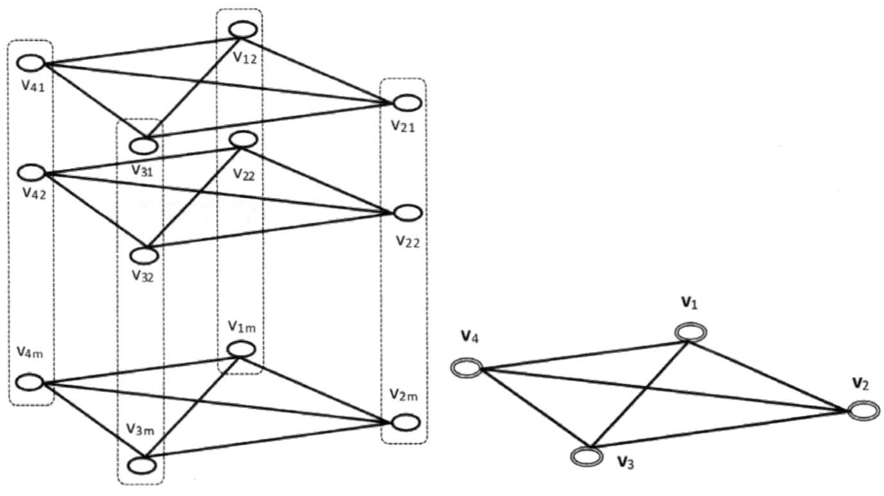

(a) (b)

Figure 3. Illustration of connections among the SH-spin vector on an Ising model: (a) SH-spin vector connections among the individual spin, and (b) each group of spins represented as a single SH-spin vectors drawn as double-lined circles, then connected as a complete graph.

Algorithm 1. H-Matrix search by simulated annealing.

Input: Order of SH matrix $4k$, annealing schedule $p_T(t)$, MaxIter
Output: A $4k$-order SH-matrix
Randomly generate $4k$-order Q-matrix Q
$t \leftarrow 0$
while $E(Q) > 0$ or $t <$ MaxIter **do**
    Copy Q into a template: $Q_t \leftarrow Q$
    Randomly select a pair of $\{+1, -1\}$ at a randomly selected column of Q_t, then flip (or exchange) the pair.
    if $E(Q_t) < E(Q)$ or random $> p_T(t)$ **then**
        $Q \leftarrow Q_t$
    end if
    $t \leftarrow t + 1$
end while

4. Experiments and analysis

In the experiment, we set the algorithm to find an H-matrix of order 12, which is the lowest order of an H-matrix that cannot be constructed by the Sylvester method. The SA schedule is displayed in figure 4(a) as a threshold function, whereas evolution of the energy curve of Q during the SA is displayed in figure 4(b). We observe fluctuation in the energy curve, indicating the Metropolis update scheme works as expected. We also find the trend of decreasing energy in figure 4(b), although it fluctuates over time; it is large in the beginning and suppressed at later iterations. The system finally reach the ground state at iteration number around $70,000$ with $E(Q) = 0$, indicating that an H-matrix (or SH-matrix) has been found.

The initial 12-order randomly generated Q is displayed in figure 5(a), where black squares indicate -1 entries, whereas the white ones correspond to $+1$. The orthogonality level among the SH vectors of Q is displayed in figure 5(b), with black indicating orthogonal, whereas other

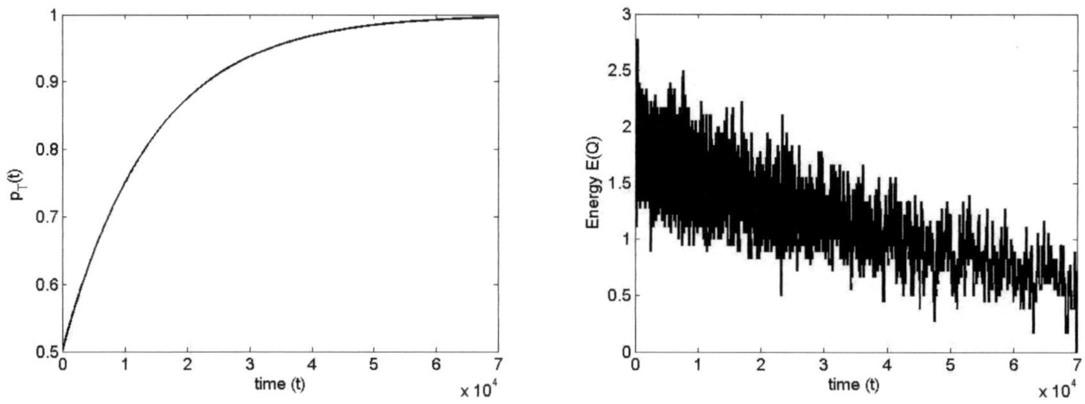

(a) (b)

Figure 4. Execution of the algorithm. Curve of (a) annealing schedule expressed in $p_T(t)$, and (b) evolution of the energy (normalized by m^2).

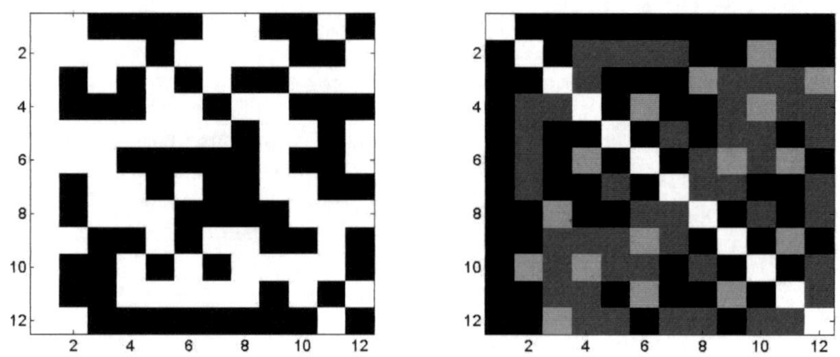

(a) (b)

Figure 5. Initial condition: (a) randomly generated Q-matrix, and (b) orthogonality indicator matrix $D = Q^TQ$

color/gray level indicating nonorthogonal vectors. After reaching the ground state, the H-matrix is found and it is displayed in figure 6(a), while orthogonality is indicated by the diagonal form of indicator matrix in figure 6(b).

Table 1 shows performance of the algorithm for the five lowest-order H-matrices, including the 12-order and 20-order that cannot be constructed by the Sylvester method. The table shows that iteration time grows nonlinearly. Taking the logarithm indicates the close relationship between $k = m/4$ with it, so we can expect that the growth is like $\mathcal{O}(e^{ck})$ for an arbitrary constant c. The column 'MaxIter' shows the numbers of maximum iteration set in the program for a given instance.

We compare the iteration number used to find the H-matrix (actually an SH-matrix) N_{SA} with the number of binary matrix $N_B(4k)$ given by (1) and the number of unique quasi H-matrix $N_{QU}(4k)$ given by (4); they are displayed in figure 7. The curves show that iteration number of finding a H-matrix grows more slowly than both of the numbers of the binary matrix and unique Q-matrix. Therefore, performing SA is much more efficient than exhaustively checking of

IOP Conf. Series: Journal of Physics: Conf. Series **856** (2017) 012012 doi:10.1088/1742-6596/856/1/012012

Table 1. Performance of the proposed method.

k	Order	NIter	log(NIter)	MaxIter
1	4	9	0.95	16,000
2	8	4,637	3.67	64,000
3	12	69,900	4.84	144,000
4	16	1,551,886	6.19	2,560,000
5	20	29,548,458	7.47	40,000,000

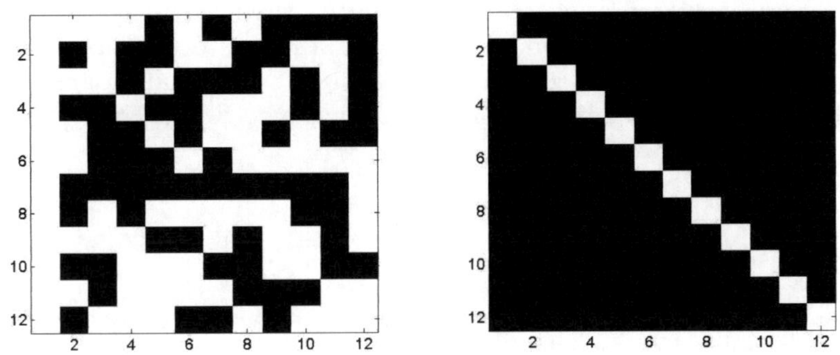

(a) (b)

Figure 6. Result of the algorithm for 12-order H-matrix search: (a) H-matrix found, and (b) indicator matrix showing orthogonality in Q.

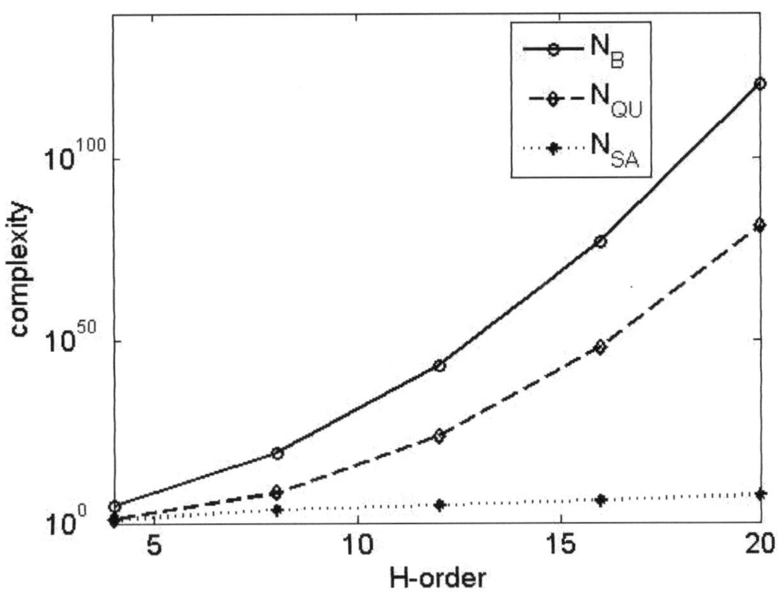

Figure 7. Comparison of N_B, N_{QU} and the number of SA iterations used to find the H-matrix N_{SA}.

either all $4k$-order binary matrix or $4k$-order Q-matrix. Nevertheless, the complexity of finding an H-matrix using SA seems to grow exponentially by $\mathcal{O}(e^{ck})$, which makes the algorithm not effective anymore for finding a high-order H-matrix. Implementing the SA code on parallel or cluster might help find a few more high order matrices, but finding the lowest order unknown H-matrix of order $4k$, i.e., 668 forces us to consider a quantum computer. The adiabatic-quantum computer with sufficient connections and number of qubits, like the D-Wave system [22], is prospective for finding such a high-order H-matrix.

5. Conclusions and further directions

We have presented a searching method for finding a Hadamard matrix using SA algorithm. Replacing the individual spins with an SH-spin vector is found effective in finding the SH matrix, which is applied successfully to find the five lowest-order H-matrices. Although the method is applicable to a general order of H-matrix, the execution time grows nonlinearly. To find high order H-matrix, we may consider a computer with multiprocessor or a cluster computer since the SA itself is, in principle, parallelizable. It is also worth to consider quantum SA algorithm to do the search and to use adiabatic-quantum computing to find such a high-order matrix in the future.

References

[1] Sylvester J J 1867 *Philos. Mag.* **34** 461
[2] Hadamard J 1893 *B. Sci. Math.* **17** 240
[3] Garg V 2007 *Wireless Communications and Networking* (Burlington: Morgan-Kaufmann)
[4] Seberry J, Wyosicki B J and Wyosicki T A 2005 *Metrika* **62** 221
[5] Paley R E A C 1933 *Journal of Mathematics and Physics* **12** 311
[6] Dade E C and Goldberg K 1959 *Mich. Math. J.* **6** 247
[7] Williamson J 1944 *Duke Math. J.* **11** 65
[8] Bush K A 1971 *J. Comb. Theory* A **11** 38
[9] Bush K A 1971 *Atti del Convegno di Geometria Combinatoria e sue Applicazioni* ed Barlotti A (Perugia) p 131
[10] Wallis J S 1976 *J. Comb. Theory* A **21** 188
[11] Karaghani H and Tayfeh-Rezaie B 2005 *J. Comb. Des.* **13** 435
[12] Kirkpatrick S, Gellat Jr C D and Vecchi M P 1983 *Science* **220** 671
[13] Černý V 1983 *J. Optimiz. Theory App.* **45** 41
[14] Metropolis N, Rosenbluth A W, Rosenbluth M N and Teller A H 1953 *J. Chem. Phys.* **21** 1087
[15] Geman S and Geman D 1984 *IEEE T. Pattern Anal.* **PAMI-6** 721
[16] Suksmono A B and Hirose A 2002 *IEEE T. Signal Proces.* **50** 290
[17] Newman M E J and Barkema G T 1999 *Monte Carlo Methods in Statistical Physics* (Oxford: Clarendon Press)
[18] Peierls R 1936 *Math. Proc. Cambridge* **32** 477
[19] Onsager L 1944 *Phys. Rev.* **65** 117
[20] Baxter R J 1982 *Exactly Solved Models in Statistical Mechanics* (San Diego: Academic Press)
[21] De las Cuevas G and Cubitt T S 2016 *Science* **351** 1180
[22] Johnson M W *et al* 2011 *Nature* **473** 194

IOP Conf. Series: Journal of Physics: Conf. Series **856** (2017) 012013 doi:10.1088/1742-6596/856/1/012013

On the internal soliton propagation over slope-shelf topography

Albert Sulaiman

Geostech Laboratory 820, Badan Pengkajian dan Penerapan Teknologi (BPPT), Kawasan Puspiptek Serpong, Tangerang Selatan 15314, Indonesia

E-mail: albertus.sulaiman@bppt.go.id

Abstract. Dynamics and properties of internal soliton propagation over slope-shelf topography is investigated. We derive a nonlinear wave equation based on the two-layer fluid model, which produce a variable-coefficient perturbed Korteweg-deVries (vP-KdV) equation. A special solution in term of single-soliton will be highlighted.

1. Introduction

Internal wave transmits energy to the ocean interior up to the sea floor. Like waves of sea surface, the internal waves will experience a transformation of energy at a time when approaching the beach. The transformation of energy depends on the sea bottom topography. The depth of the sea will change from deep to shallow with varied forms. The most common is in the shape of a slope-shelf.

The propagation of internal wave, especially internal solitary waves (ISW) or internal soliton, from shelf break has remarkable properties. The internal waves would propagate into much shallower water and transform to the elevation waves [1–3]. The ISW propagation that usually was described by Korteweg de Vries (KdV) equation will change sign in its coefficient. The comparison of observation by using satellite image and in-situ experiment showed the sign is negative in deep water and positive in shallow water. The wave transform from waves of depression to waves of elevation, for example the depression propagating from a deep part of a basin onto the shelf, will break when the wave amplitude is larger than a half of the water depth minus the undisturbed depth of the isopycnal of maximum depression [4–7].

In this paper, we will develop an analytic model of the ISW propagation over slope-shelf topography. The two-layered fluid model will be used to derive KdV form. The special solution will be used to examine the nature of dynamics of ISW such as run-up and the breaking waves.

2. Derivation of nonlinear internal waves propagation over slope-shelf topography

The derivation of ISW propagation over uneven bottom has been done [4]. In this paper, we will derive another approach and show the limitation of previous derivation. We consider a two-layer fluid bounded above by a rigid horizontal plane (the rigid-lid approximation), $z = h_1$ and below by a rigid horizontally-varying boundary $z = -h_2(x)$. Each layer consists of incompressible, inviscid and irrotational fluid of a constant density ρ_1 for the upper layer and ρ_2 for the lower layer. The free interface between the layer is denoted by $z = \eta(x, t)$. The geometry of boundary

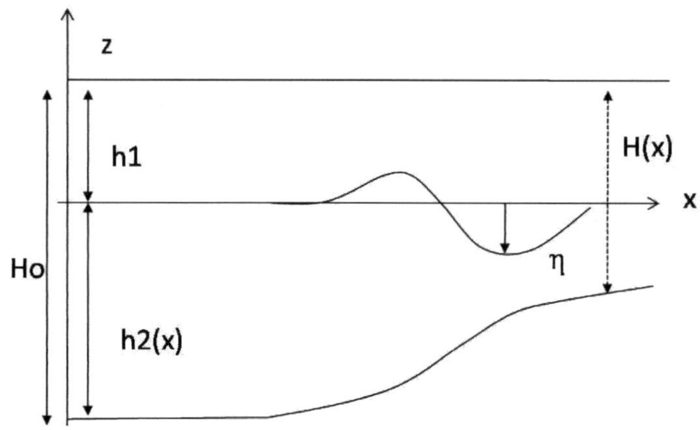

Figure 1. The boundary problem of two-layer fluid with rigid lid on the upper layer and uneven bottom on the lower layer.

value problems is depicted in figure 1. The derivation of internal soliton wave equation will follow procedures as discussed by [8].

We begin with the equations of motion

$$\frac{\partial u_i}{\partial t} + u_i\frac{\partial u_i}{\partial x} + w_i\frac{\partial u_i}{\partial z} = -\frac{1}{\rho_0}\frac{\partial p_i}{\partial x}, \tag{1}$$

$$\frac{\partial w_i}{\partial t} + u_i\frac{\partial w_i}{\partial x} + w_i\frac{\partial w_i}{\partial z} = -\frac{1}{\rho_0}\frac{\partial p_i}{\partial z}, \tag{2}$$

$$\frac{\partial u_i}{\partial x} + \frac{\partial w_i}{\partial z} = 0, \tag{3}$$

and their boundary values are given by

$$w_1|_{z=h_1} = 0 \tag{4}$$

$$w_2|_{z=-h_2} = -u_2\frac{\partial h_2(x)}{\partial x} \tag{5}$$

$$w_i|_{z=\eta} = \frac{\partial \eta}{\partial t} + u_i\frac{\partial \eta}{\partial x} \tag{6}$$

$$(p_2 - p_1)|_{z=\eta} = (\rho_2 - \rho_1)\, g\eta \tag{7}$$

where u_i, w_i are orbital velocities for horizontal and vertical respectively, and ρ_i is the fluid density; here index i takes the value 1 for the upper layer and 2 for the lower layer. Now, we introduce a scaling as follows: $[t] = T$, $[x] = L$, $[z] = H_0$, $[u] = \epsilon c_0$, $[w] = \epsilon c_0 H_0/L$, $[\eta] = \epsilon H_0$ and $[p] = \rho_0\epsilon c_0^2$, where ϵ is a small parameter and c is phase velocity. The phase speed is given by $c_0^2 = \bar{g}H_0$ with $\bar{g} = g(\rho_2 - \rho_1)/\rho_2$ and $H_0 = h_1 + h_2$ for constant h_2.

Using these scales, we can cast (1) and (2) in a nondimensional form

$$\frac{\partial u_i}{\partial t} + \epsilon u_i\frac{\partial u_i}{\partial x} + \epsilon w_i\frac{\partial u_i}{\partial z} = -\frac{\partial p_i}{\partial x} \tag{8}$$

$$\frac{\partial w_i}{\partial t} + \epsilon u_i\frac{\partial w_i}{\partial x} + \epsilon w_i\frac{\partial w_i}{\partial z} = -\frac{1}{\delta}\frac{\partial p_i}{\partial z} \tag{9}$$

$$\frac{\partial u_i}{\partial x} + \frac{\partial w_i}{\partial z} = 0 \tag{10}$$

IOP Conf. Series: Journal of Physics: Conf. Series **856** (2017) 012013 doi:10.1088/1742-6596/856/1/012013

where $\delta = H_0^2/L^2$. The boundary conditions can be written in the nondimensional form,

$$
\begin{aligned}
w_1|_{z=\alpha} &= 0 && (11)\\
w_2|_{z=(\alpha-1)\bar{H}} &= -u_2\frac{\partial \bar{h}_2(x)}{\partial x} && (12)\\
w_i|_{z=\epsilon\eta} &= \frac{\partial \eta}{\partial t} + \epsilon u_i \frac{\partial \eta}{\partial x} && (13)\\
(p_2 - p_1)|_{z=\epsilon\eta} &= \eta && (14)
\end{aligned}
$$

where $h_1 = \alpha\bar{H}$, $\bar{h}_2(x) = (1-\alpha)\bar{H}$, $\bar{H} = H/H_0$ and $H = H(x) = h_1 + h_2(x)$. Furthermore, we assume that $\epsilon = \mathcal{O}(\delta) \ll 1$ and $\delta = \epsilon\delta_*$ with $\delta_* = \mathcal{O}(1)$ [8]. By using (8)–(14), we will obtain the dynamics of internal wave η describing by

$$D_0(\eta) + \epsilon D_1(\eta) + (\epsilon^2) = 0\ , \qquad (15)$$

where D_0 and D_1 are differential operators.

The problem to be solved is the boundary condition at $z = \eta$. It means that we still do not know the position of water particle at the interface. This can be obtained by applying the Taylor expansion of each term that depends on z around $z = 0$. Applying the Taylor expansion to (13) and using the continuity equation yield

$$w_i|_{z=\epsilon\eta} = w_i|_{z=0} - \epsilon\eta\frac{\partial u_i}{\partial x} + \mathcal{O}(\epsilon^2)\ . \qquad (16)$$

Comparing with (13) we have

$$w_i|_{z=0} = \frac{\partial \eta}{\partial t} + \epsilon\frac{\partial(\eta u_i)}{\partial x} + \mathcal{O}(\epsilon^2)\ . \qquad (17)$$

The boundary condition (14) can be written as

$$\eta = (p_2 - p_1)|_{z=0} + \epsilon\eta\left(\frac{\partial p_2}{\partial z} - \frac{\partial p_1}{\partial z}\right)_{z=0} + \mathcal{O}(\epsilon^2)\ . \qquad (18)$$

We also define a shear cross-interface as $\bar{u} = (u_2 - u_1)|_{z=\epsilon\eta}$ and its Taylor expansion as follows

$$\bar{u} = (u_2 - u_1)|_{z=0} + \epsilon\eta\left(\frac{\partial u_2}{\partial z} - \frac{\partial u_1}{\partial z}\right)_{z=0} + \mathcal{O}(\epsilon^2)\ . \qquad (19)$$

We develop variables p_i, u_i, w_i and η in a series, in which ϵ serves as small parameter:

$$\begin{pmatrix} u_i \\ w_i \\ \eta_i \\ p_i \end{pmatrix} = \begin{pmatrix} u_i^{(0)} \\ w_i^{(0)} \\ \eta^{(0)} \\ p_i^{(0)} \end{pmatrix} + \epsilon \begin{pmatrix} u_i^{(1)} \\ w_i^{(1)} \\ \eta^{(1)} \\ p_i^{(1)} \end{pmatrix} + \mathcal{O}(\epsilon^2)\ . \qquad (20)$$

By using this expansion, the equations of motion (8), (9) and (10) become

$$
\begin{aligned}
&\frac{\partial u_i^{(0)}}{\partial t} + \frac{\partial p_i^{(0)}}{\partial x} + \epsilon\frac{\partial u_i^{(1)}}{\partial t} + \epsilon u_i^{(0)}\frac{\partial u_i^{(0)}}{\partial x} + \epsilon w_i^{(0)}\frac{\partial u_i^{(0)}}{\partial x} + \epsilon\frac{\partial p_i^{(1)}}{\partial z} + \mathcal{O}(\epsilon^2) = 0, && (21)\\
&\frac{\partial p_i^{(0)}}{\partial z} + \epsilon\delta_*\frac{\partial w_i^{(0)}}{\partial t} + \epsilon\frac{\partial p_i^{(1)}}{\partial z} + \mathcal{O}(\epsilon^2) = 0, && (22)\\
&\frac{\partial u_i^{(0)}}{\partial x} + \frac{\partial w_i^{(0)}}{\partial z} + \epsilon\left(\frac{\partial u_i^{(1)}}{\partial x} + \frac{\partial w_i^{(1)}}{\partial z}\right) + \mathcal{O}(\epsilon^2) = 0\ , && (23)
\end{aligned}
$$

IOP Conf. Series: Journal of Physics: Conf. Series **856** (2017) 012013 doi:10.1088/1742-6596/856/1/012013

and the boundary conditions (11), (12), (17), (27) and (19) give

$$w_1^{(0)}|_{z=\alpha} + \epsilon w_1^{(1)}|_{z=\alpha} + \mathcal{O}(\epsilon^2) = 0, \tag{24}$$

$$w_2^{(0)}|_{z=(\alpha-1)\bar{H}} + \epsilon w_2^{(1)}|_{z=(\alpha-1)\bar{H}} + \epsilon u_2^{(0)} \frac{\partial \bar{h}_2(x)}{\partial x} + \mathcal{O}(\epsilon^2) = 0, \tag{25}$$

$$w_i^{(0)}|_{z=0} + \epsilon w_i^{(1)}|_{z=0} - \frac{\partial \eta^{(0)}}{\partial t} - \epsilon \frac{\partial \eta^{(1)}}{\partial t} - \epsilon \frac{\partial \left(\eta^{(0)} u_i^{(0)}\right)}{\partial x} + \mathcal{O}(\epsilon^2) = 0, \tag{26}$$

$$\eta^{(0)} + \epsilon \eta^{(1)} = \left(p_2^{(0)} - p_1^{(0)}\right)\Big|_{z=0} + \epsilon \left(p_2^{(1)} - p_1^{(1)}\right)\Big|_{z=0} + \epsilon \eta^{(0)} \left(\frac{\partial p_2^{(0)}}{\partial z} - \frac{\partial p_1^{(0)}}{\partial z}\right)_{z=0} + \mathcal{O}(\epsilon^2), \tag{27}$$

$$\bar{u}^{(0)} + \epsilon \bar{u}^{(1)} = \left(u_2^{(0)} - u_1^{(0)}\right)\Big|_{z=0} + \epsilon \left(u_2^{(1)} - u_1^{(1)}\right)\Big|_{z=0} + \epsilon \eta^{(0)} \left(\frac{\partial u_2^{(0)}}{\partial z} - \frac{\partial u_1^{(0)}}{\partial z}\right)_{z=0} + \mathcal{O}(\epsilon^2)\,, \tag{28}$$

where we have assumed that the variation of bottom topography acts as a perturbation of solitary wave, which means $\bar{h}_2(x) = \epsilon \bar{h}_2(x)$.

At the lowest order, $\epsilon^{(0)}$, the equations of motion and their boundary conditions read

$$\frac{\partial u_i^{(0)}}{\partial t} + \frac{\partial p_i^{(0)}}{\partial x} = 0, \quad \frac{\partial p_i^{(0)}}{\partial z} = 0, \quad \frac{\partial u_i^{(0)}}{\partial x} + \frac{\partial w_i^{(0)}}{\partial z} = 0, \quad w_1^{(0)}|_{z=\alpha} = 0, \quad w_2^{(0)}|_{z=(\alpha-1)\bar{H}} = 0, \tag{29}$$

$$w_i^{(0)}|_{z=0} - \frac{\partial \eta^{(0)}}{\partial t} = 0, \quad \eta^{(0)} = \left(p_2^{(0)} - p_1^{(0)}\right)|_{z=0}, \quad \bar{u}^{(0)} = \left(u_2^{(0)} - u_1^{(0)}\right)|_{z=0}\,. \tag{30}$$

We see that, in the lowest order, $u_i^{(0)}$ and $p_i^{(0)}$ do not depend on vertical coordinate z. Writing down the first and the second of (29) into their components, we get

$$\frac{\partial u_1^{(0)}}{\partial t} + \frac{\partial p_1^{(0)}}{\partial x} = 0, \quad \frac{\partial u_2^{(0)}}{\partial t} + \frac{\partial p_2^{(0)}}{\partial x} = 0 \tag{31}$$

Subtracting the two equations yields

$$\frac{\partial \bar{u}^{(0)}}{\partial t} + \frac{\partial \eta^{(0)}}{\partial x} = 0\,. \tag{32}$$

Next, we integrate the continuity equation, i.e, the third equation of (29), with respect to vertical coordinate with the upper and lower bounds. For the upper layer,

$$\frac{\partial u_1^{(0)}}{\partial x}(\alpha - 0) + w_1^{(0)}\,|_{z=\alpha} - w_1^{(0)}\,|_{z=0} = 0\,. \tag{33}$$

By using the boundary condition, we get

$$\alpha \frac{\partial u_1^{(0)}}{\partial x} - \frac{\partial \eta^{(0)}}{\partial t} = 0\,. \tag{34}$$

For the lower layer,

$$\frac{\partial u_2^{(0)}}{\partial x}\left(0 - (\alpha - 1)\bar{H}\right) + w_2^{(0)}\,|_{z=0} - w_2^{(0)}\,|_{z=(\alpha-1)\bar{H}} = 0\,. \tag{35}$$

By using the boundary condition and multiplying (34) by $(\alpha-1)\bar{H}$, and adding up the results we get

$$\alpha(\alpha-1)\bar{H}\frac{\partial \bar{u}^{(0)}}{\partial x}+\left((\alpha-1)\bar{H}-\alpha\right)\frac{\partial \eta^{(0)}}{\partial t}=0\,. \tag{36}$$

Finally, by using (34), we get

$$\frac{\partial^2 \eta^{(0)}}{\partial t^2}-c_0^2\frac{\partial^2 \eta^{(0)}}{\partial x^2}=0\,, \tag{37}$$

where $c_0^2=(\alpha(\alpha-1)\bar{H})/(\alpha-1)\bar{H}-\alpha)$. When $h_2(x)=h_2$ with h_2 being constant and $\bar{H}=1$, we have the phase speed of linear long wave for two-fluid system. It is well known that (37) describes linear long waves, with both left and right wards propagating. Its general solution can be written as $\eta^{(0)}=F(\xi_-)+G(\xi_+)$, where $\xi_-=x-c_0t$ represents the right propagation, and $\xi_+=x+c_0t$, the left propagation.

Now, we will arrive at the crucial blackpoint, namely the first order. At the first order, $\epsilon^{(1)}$, we have

$$\frac{\partial u_i^{(1)}}{\partial t}+u_i^{(0)}\frac{\partial u_i^{(0)}}{\partial x}+\frac{\partial p_i^{(1)}}{\partial x}=0 \tag{38}$$

$$\delta_*\frac{\partial w_i^{(0)}}{\partial t}+\frac{\partial p_i^{(1)}}{\partial z}=0 \tag{39}$$

$$\frac{\partial u_i^{(1)}}{\partial x}+\frac{\partial w_i^{(1)}}{\partial z}=0\,, \tag{40}$$

for the equations of motion (note that we have used the fact that $\partial(u_i^{(0)},p_i^{(0)})/\partial z=0$) and

$$w_1^{(1)}|_{z=\alpha} = 0,\ w_2^{(1)}|_{z=(\alpha-1)\bar{H}}=-u_2^{(0)}\frac{\partial \bar{h}_2(x)}{\partial x} \tag{41}$$

$$w_i^{(1)}|_{z=0} = \frac{\partial \eta^{(1)}}{\partial t}+\frac{\partial\left(\eta^{(0)}u_i^{(0)}\right)}{\partial x} \tag{42}$$

$$\eta^{(1)} = \left(p_2^{(1)}-p_1^{(1)}\right)|_{z=0},\ \bar{u}^{(1)}=\left(u_2^{(1)}-u_1^{(1)}\right)|_{z=0}\,, \tag{43}$$

for the boundary conditions. Before we work further with the first order, we require the following relations

$$u_1^{(0)}=\frac{(1-\alpha)\bar{H}}{(\alpha-1)\bar{H}-\alpha}\bar{u}^{(0)},\quad u_2^{(0)}=\frac{\alpha}{(1-\alpha)\bar{H}+\alpha}\bar{u}^{(0)}\,. \tag{44}$$

These relations can be obtained from the zeroth order. By doing integration from α to z and using (34) for the upper layer, and similarly doing integration from $(\alpha-1)\bar{H}$ to z and using boundary condition of the zeroth order for the lower layer, we get

$$w_1^{(0)}(z)=\left(\frac{\alpha-z}{\alpha}\right)\frac{\partial \eta^{(0)}}{\partial t},\quad w_2^{(0)}(z)=\left(\frac{(\alpha-1)\bar{H}-z}{(\alpha-1)\bar{H}}\right)\frac{\partial \eta^{(0)}}{\partial t}\,. \tag{45}$$

We write the horizontal momentum equation, (38), for each layer, and then subtract and use boundary condition of the first order and (44), we get

$$\frac{\partial \bar{u}^{(1)}}{\partial t}+\Lambda\bar{u}^{(0)}\frac{\partial \bar{u}^{(0)}}{\partial x}+\frac{\partial \eta^{(1)}}{\partial x}=0\,, \tag{46}$$

where $\Lambda = (\alpha^2 - (\alpha-1)^2\bar{H}^2)/((\alpha-1)\bar{H}-\alpha)^2$. Taking differential operation with respect to z for the horizontal momentum equation and to x for the vertical momentum equation, followed by subtracting these results and taking integration with respect to t yield

$$\delta_* \frac{\partial w_i^{(0)}}{\partial x} = \frac{\partial \bar{u}^{(1)}}{\partial z} \tag{47}$$

By substituting (45) and taking integral operation with respect to z, we obtain

$$u_1^{(1)} = \frac{\delta_*}{\alpha}\left(\alpha z - \frac{1}{2}z^2\right)\frac{\partial^2\eta^{(0)}}{\partial x\partial t} + u_1^{(1)}\,|_{z=0} \tag{48}$$

$$u_2^{(1)} = \frac{\delta_*}{(\alpha-1)\bar{H}}\left((\alpha-1)\bar{H}z - \frac{1}{2}z^2\right)\frac{\partial^2\eta^{(0)}}{\partial x\partial t} + u_2^{(1)}\,|_{z=0}\ . \tag{49}$$

Same as the lowest order, integrating the continuity equation (40) with respect to z and using (48) give

$$\frac{\delta_*\alpha^2}{3}\frac{\partial^3\eta^{(0)}}{\partial x^2\partial t} + \alpha\frac{\partial u_1^{(1)}}{\partial x}\,|_{z=0} + w_1^{(1)}\,|_{z=\alpha} - w_1^{(1)}\,|_{z=0} = 0\ . \tag{50}$$

By using the first order of boundary condition and (42), we have

$$\frac{\partial\eta^{(1)}}{\partial t} + \frac{\partial\left(\eta^{(0)}u_1^{(0)}\right)}{\partial x} - \alpha\frac{\partial u_1^{(1)}}{\partial x}\,|_{z=0} - \frac{1}{3}\delta_*\alpha^2\frac{\partial^3\eta^{(0)}}{\partial x^2\partial t} = 0\ . \tag{51}$$

In the same manner, we get the lower layer as follows

$$-\frac{\partial}{\partial x}\left[\frac{\delta_*}{3}(\alpha-1)^2\bar{H}^2\frac{\partial^2\eta^{(0)}}{\partial x\partial t} + (\alpha-1)\bar{H}u_2^{(1)}\,|_{z=0}\right] + w_2^{(1)}\,|_{z=0} - w_2^{(1)}\,|_{z=(\alpha-1)\bar{H}} = 0\ .$$

By using (41) and (42), we arrive at

$$\begin{aligned}&\frac{\partial\eta^{(1)}}{\partial t} + \frac{\partial}{\partial x}\left(\eta^{(0)}u_2^{(0)}\right) + u_2^{(0)}\frac{\partial h_2}{\partial x} + (1-\alpha)\frac{\partial\bar{H}}{\partial x}u_2^{(1)}\,|_{z=0} + (1-\alpha)\bar{H}\frac{\partial u_2^{(1)}}{\partial x}\,|_{z=0}\\ &-\frac{\delta_*}{3}(\alpha-1)^2\frac{\partial\bar{H}^2}{\partial x}\frac{\partial^2\eta^{(0)}}{\partial x\partial t} - \frac{\delta_*}{3}(\alpha-1)^2\bar{H}^2\frac{\partial^3\eta^{(0)}}{\partial x^2\partial t} = 0\ .\end{aligned} \tag{52}$$

By assuming that $H \gg h_2$ and considering that the varying depth is a perturbation of internal waves, then we can ignore the derivative of H; thus by substituting (44), we get

$$\frac{\partial\eta^{(1)}}{\partial t} + \frac{\partial}{\partial x}\left(\eta^{(0)}u_2^{(0)}\right) + \frac{\alpha}{(1-\alpha)\bar{H}+\alpha}\bar{u}^{(0)}\frac{\partial h_2}{\partial x} + (1-\alpha)\bar{H}\frac{\partial u_2^{(1)}}{\partial x}\,|_{z=0} - \frac{\delta_*}{3}(\alpha-1)^2\bar{H}^2\frac{\partial^3\eta^{(0)}}{\partial x^2\partial t} = 0\ . \tag{53}$$

Multiplying (51) by $(1-\alpha)\bar{H}$, (53) with α, followed by adding them up and using (43), we obtain

$$\begin{aligned}&(\alpha+(1-\alpha)\bar{H})\frac{\partial\eta^{(1)}}{\partial t} + \alpha(1-\alpha)\bar{H}\frac{\partial\bar{u}^{(1)}}{\partial x} + \frac{\alpha^2}{(1-\alpha)\bar{H}+\alpha}\bar{u}^{(0)}\frac{\partial h_2}{\partial x}\\ &+\alpha\frac{\partial\left(u_2^{(0)}\eta^{(0)}\right)}{\partial x} + (1-\alpha)\bar{H}\frac{\partial\left(\bar{u}_1^{(0)}\eta^{(0)}\right)}{\partial x} - \frac{\delta_*}{3}\alpha\bar{H}\left(\alpha+(1-2\alpha)\bar{H}\right)\frac{\partial^3\eta^{(0)}}{\partial x^2\partial t} = 0\ .\end{aligned} \tag{54}$$

Multiplying by -1 and using the boundary condition of the first order and assuming the second derivative of h_2 can be ignored, we then get

$$\frac{\partial^2\eta^{(1)}}{\partial t^2} - c_0^2\frac{\partial^2\eta^{(1)}}{\partial x^2} + 2\Lambda\frac{\partial}{\partial x}\left(\eta^{(0)}\frac{\partial\eta^{(0)}}{\partial x}\right) - \Delta_1\frac{\partial}{\partial x}\left(\frac{\partial h_2}{\partial x}\eta^{(0)}\right) + \Delta_2\frac{\partial^4\eta^{(0)}}{\partial x^4} = 0 , \quad (55)$$

where $\Delta_1 = \alpha^2/((\alpha-1)\bar{H}-\alpha)^2$ and $\Delta_2 = (\delta_*\alpha\bar{H}(\alpha+(1-2\alpha)\bar{H})c_0^2)/(3((\alpha-1)\bar{H}-\alpha))$. Because the function of $\eta^{(0)}$ also meets the wave equation (37), then the derivative with respect to x can be split as follows

$$\frac{\partial}{\partial x}\eta^{(0)} = -\frac{1}{2c_0}\left(\frac{\partial}{\partial t} - c_0\frac{\partial}{\partial x}\right)\eta^{(0)} . \quad (56)$$

Using the operator we write

$$\left(\frac{\partial}{\partial t} - c_0\frac{\partial}{\partial x}\right)\left(\frac{\partial}{\partial t} + c_0\frac{\partial}{\partial x}\right)\eta^{(1)} - \frac{\Lambda}{c_0}\left(\frac{\partial}{\partial t} - c_0\frac{\partial}{\partial x}\right)\left(\eta^{(0)}\frac{\partial\eta^{(0)}}{\partial x}\right)$$
$$-\frac{\Delta_1}{2c_0}\left(\frac{\partial}{\partial t} - c_0\frac{\partial}{\partial x}\right)\left(\frac{\partial h_2}{\partial x}\eta^{(0)}\right) - \frac{\Delta_2}{2c_0}\left(\frac{\partial}{\partial t} - c_0\frac{\partial}{\partial x}\right)\frac{\partial^3\eta^{(0)}}{\partial x^3} = 0 , \quad (57)$$

with integration constant being zero; this yields

$$\left(\frac{\partial}{\partial t} + c_0\frac{\partial}{\partial x}\right)\eta^{(1)} - \frac{\Lambda}{c_0}\eta^{(0)}\frac{\partial\eta^{(0)}}{\partial x} - \frac{\Delta_1}{2c_0}\frac{\partial h_2}{\partial x}\eta^{(0)} - \frac{\Delta_2}{2c_0}\frac{\partial^3\eta^{(0)}}{\partial x^3} = 0 . \quad (58)$$

Finally, after thorough effort, we get evolution equation for η as

$$\frac{\partial\eta}{\partial t} + c_0\frac{\partial\eta}{\partial x} - \frac{\Lambda}{c_0}\eta\frac{\partial\eta}{\partial x} - \frac{\Delta_2}{2c_0}\frac{\partial^3\eta}{\partial x^3} - \frac{\Delta_1}{2c_0}\frac{\partial h_2}{\partial x}\eta = 0 . \quad (59)$$

This is called the variable-coefficient perturbed Korteweg-deVries (vP-KdV) equation.

3. Special solution of ISW propagation over slope-shelf topography

Before we solve (59), let us write the equation in the form of

$$\frac{\partial\eta}{\partial t} + c_0\frac{\partial\eta}{\partial x} + \gamma_1\eta\frac{\partial\eta}{\partial x} + \gamma_2\frac{\partial^3\eta}{\partial x^3} + F(x)\eta = 0 , \quad (60)$$

where $\gamma_1 = -\Lambda/c_0$, $\gamma_2 = -\Delta_2/(2c_0)$ and $F(x) = -\Delta_1/(2c_0)(\partial h_2/\partial x)$. The single-soliton solution of the equation is well-known when $F(x) = 0$ and the coefficients are constant. It has been shown that by using transformation

$$\psi = \exp[\int \frac{2F(x')}{c_0}dx']\eta, \tau = \int^x \frac{dx'}{c_0}, \varsigma = \tau - t , \quad (61)$$

equation (60) reduces to [9]

$$\frac{\partial\psi}{\partial\tau} + \delta_1(\tau)\psi\frac{\partial\psi}{\partial\varsigma} + \delta_2(\tau)\frac{\partial^3\psi}{\partial\varsigma^3} = 0 , \quad (62)$$

where $\delta_1 = \gamma_1/(c_0\exp[\int 2F/c_0 dx]$ and $\delta_2 = \gamma_2/c_0^3$. Now, the coefficients are only function of τ alone. An analytic solution based on slowly varying topography approximation has been done by several researcher [9]. The slowly varying topography means that the topography undulation is larger than solitary wave wavelength. The solution can be expressed as follows

$$\psi = \psi_0(\bar{\tau}, \bar{\varsigma}) + \epsilon\psi_1(\bar{\tau}, \bar{\varsigma}) + o(\epsilon^2) \ , \tag{63}$$

where $\bar{\tau} = \epsilon\tau$ and $\bar{\varsigma} = \epsilon\varsigma$. The zeroth order leads to single-soliton solution given by [4]

$$\psi_0(\bar{\varsigma}, \bar{\tau}) = A_0 \operatorname{sech}^2 [\gamma(\bar{\varsigma} - V_0\bar{\tau})] \ , \tag{64}$$

where $A_0 = 3V_0/\delta_1$ and $\gamma = \sqrt{\delta_1 A_0/(12\delta_2)}$. The first order satisfies the third-order ordinary differential equation with ψ_0 coefficient. By using the conservation of *mass* and *momentum*, the solution is given by

$$\psi_1(\bar{\varsigma}, \bar{\tau}) = A_1 \operatorname{sech}^2 \left[\sqrt{\frac{A_1}{2}}\,(\bar{\varsigma} - V_1\bar{\tau})\right] , \tag{65}$$

where $A_1 = \sqrt{V_1^2/4 + 2A_0 + 2A_0 + A_0^2} - V_0^2/2$ and $V_1 = 2A_0 + 2A_1$. Another way to find the analytic solution of variable-coefficient KdV equation has been developed by expansion methods based on the Riccati ordinary differential equation. The solution is given by [10]

$$\psi(x, t) = \psi_0 sech \left[\gamma \left(\varsigma - V(\tau)\tau\right)\right] \ , \tag{66}$$

where

$$V(\tau) = \frac{1}{\tau}\left(\int \delta_1(\tau) \int \delta_2(\tau')d\tau' d\tau + \int \frac{\psi_0\delta_1}{3} d\tau\right) \ , \tag{67}$$

With the special form of the topography, then solution could be obtained.

4. Conclusion

Propagation of internal solitary wave in two layer fluid have been derived. We obtain the perturbed variable coefficient Kortweg and de Vries equation. By usung slowly varying bottom topography the soliton solution can be obtained.

Acknowledgments

This work is partially funded by DIPA PTPSW-BPPT for fiscal year 2016, Risetpro 2013 and Pauli Center for Theoretical Studies, ETH Zurich 2013.

References

[1] Fu K H, Wang Y H, Laurent L S, Simmons H and Wang D P 2012 *Cont. Shelf Res.* **37** 1
[2] Holloway P E, Pelinovsky E, Talipova T and Barnes B 1997 *J. Phys. Oceanogr.* **27** 871
[3] Pritchard M and Weller R A 2005 *J. Geophys. Res. C: Oceans* **110** C03020
[4] Helfrich K R and Melville W K 2006 *Annu. Rev. Fluid Mech.* **38** 395
[5] Shishkina O D, Sveen J K and Grue J 2013 *Nonlinear Processes Geophys.* **20** 743
[6] Helfrich K R and Melville W K 1986 *J. Fluid Mech.* **167** 285
[7] Vlasenko V and Hutter K 2002 *J. Phys. Oceanogr.* **32** 1779
[8] Gerkema T and Zimmerman J T F 2008 An introduction to internal waves lecture notes Royal Netherlands Institute for Sea Research, Texel
[9] Grimshaw R 2007 *Solitary Waves Propagating over Variable Topography* (Berlin, Heidelberg: Springer Berlin Heidelberg) pp 51–64
[10] Latif M S A 2011 *Commun. Nonlinear Sci. Numer. Simul.* **16** 1783

IOP Conf. Series: Journal of Physics: Conf. Series **856** (2017) 012014 doi:10.1088/1742-6596/856/1/012014

Conservative formulation of unsteady pipe-flow model for water in its liquid form

Bambang Supriyadi[1], I Gusti Ketut Puja[2], F A Rusdi Sambada[2] and Sudi Mungkasi[3]

[1] Department of Civil and Environmental Engineering, Faculty of Engineering, Gadjah Mada University, Jalan Grafika No. 2, Yogyakarta 55281, Indonesia
[2] Department of Mechanical Engineering, Faculty of Science and Technology, Sanata Dharma University, Mrican, Tromol Pos 29, Yogyakarta 55002, Indonesia
[3] Department of Mathematics, Faculty of Science and Technology, Sanata Dharma University, Mrican, Tromol Pos 29, Yogyakarta 55002, Indonesia

E-mail: bbsupri@tsipil.ugm.ac.id, ketut@usd.ac.id, sambada@usd.ac.id, sudi@usd.ac.id

Abstract. We consider flows in a pipe system. The problem relating to the pipe system usually occurs in real life as well as in the laboratory, such as in the distillation process. This unsteady problem can be modeled into a hyperbolic system of partial differential equations. Our goal is to propose the use of a conservative finite-volume numerical method for solving the unsteady pipe-flow model. We obtain that the method is able to solve the pipe-flow problem successfully.

1. Introduction

Pipe flows occur in daily life as well as in the laboratory. In the laboratory, pipe flows can be found in the water distillation process. Distillation itself includes two types of water forms, gas and liquid.

Mathematical models for pipe flows have been available in the literature [1]. In the gaseous form, Mungkasi [2] used a conservative finite-volume numerical method for solving the Euler equations of gas dynamics. In the liquid form, Škifić *et al* [3] proposed a nonconservative formulation of unsteady pipe-flow problems. The conservative formulation for water in its liquid form was not given in the work of Škifić *et al* [3].

This paper complements the work of Škifić *et al* [3]. We propose the use of a conservative formulation of unsteady pipe-flow model for water in its liquid form. Because the model is hyperbolic, we choose a finite-volume method as our conservative formulation to solve the model. Finite-volume methods have been well known as powerful methods to reconstruct smooth and nonsmooth functions as solutions to problems [4].

This paper is organized as follows. We recall the mathematical model governing the problem in Section 2. We write the finite-volume scheme to solve the model in Section 3. Numerical results are provided and discussed in Section 4. Some remarks conclude the paper in Section 5.

IOP Conf. Series: Journal of Physics: Conf. Series **856** (2017) 012014 doi:10.1088/1742-6596/856/1/012014

2. Mathematical model

In this section, we write the mathematical model and the general formulation for solving the model. We assume that the model is one-dimensional.

The mathematical model governing conservation laws of water flowing in a pipe [3] is

$$H_t + \frac{c^2}{gA} Q_x = 0, \tag{1}$$

$$Q_t + gAH_x = 0. \tag{2}$$

Here, the involved variables and parameters are as follows: t is the time variable, x is the space variable for coordinate along the conduit length, $H = H(x,t)$ denotes the piezometric head, $Q = Q(x,t)$ represents the discharge, A means the pipe cross-sectional area, c is the wave speed, and g is the acceleration due to gravity. In this paper, we assume that there is no head losses in the system.

Equations (1) and (2) can be written in the vector form of conservation laws

$$\mathbf{u}_t + \mathbf{f}(\mathbf{u})_x = \mathbf{0}, \tag{3}$$

where the vector of conserved quantities is

$$\mathbf{u} = \begin{pmatrix} H \\ Q \end{pmatrix} \tag{4}$$

and the vector of flux functions is

$$\mathbf{f}(\mathbf{u}) = \begin{pmatrix} \frac{c^2}{gA} Q \\ gAH \end{pmatrix}. \tag{5}$$

The Jacobian matrix of the vector of flux functions is defined as

$$\mathbf{A}(\mathbf{u}) = \frac{\partial \mathbf{f}}{\partial \mathbf{u}} = \begin{pmatrix} 0 & \frac{c^2}{gA} \\ gA & 0 \end{pmatrix}, \tag{6}$$

with eigenvalues and right eigenvalue vectors are

$$\lambda^{(1),(2)} = \mp c \tag{7}$$

and

$$r^{(1),(2)} = \begin{pmatrix} \frac{\lambda^{(1),(2)}}{Ag} \\ 1 \end{pmatrix}. \tag{8}$$

Now, suppose that we have the scalar conservation law of (3), that is

$$u_t + f(u)_x = 0, \tag{9}$$

where u is a scalar (not a vector). The corresponding fully discrete finite-volume numerical scheme is

$$Q_i^{n+1} = Q_i^n - \frac{\Delta t}{\Delta x}(F_{i+1/2}^n - F_{i-1/2}^n), \tag{10}$$

where Δx is the cell width, and Δt is the time step in a finite-volume framework. Here, U_i^n is the finite-volume average approximation of the quantity at the ith cell at the nth time step, and $F_{i+1/2}^n$ is the finite-volume average approximation of the flux function at the $(i+1/2)$ interface from the nth time step to the $(n+1)$th time step. The scalar finite-volume scheme (10) can be extended for the vector form (3) in order to solve (1) and (2). We discuss this in the next section.

IOP Conf. Series: Journal of Physics: Conf. Series **856** (2017) 012014 doi:10.1088/1742-6596/856/1/012014

3. Numerical method

As follows, we briefly describe the numerical method that we use to solve (1) and (2). The method is a finite-volume method, which is conservative.

Let us consider (3). Using the standard finite-volume framework, let Δx be the cell width of uniformly discretized space domain, and Δt be a given time step. In the semidiscrete form, the finite-volume method is

$$\frac{d}{dt}\mathbf{U}_j(t) = -\frac{1}{\Delta x}(\mathbf{F}_{j+\frac{1}{2}}(t) - \mathbf{F}_{j-\frac{1}{2}}(t)) \tag{11}$$

where $\mathbf{U}$ is the vector of approximate conserved quantities and $\mathbf{F}$ is the vector of approximate fluxes. This scheme is called semidiscrete because we have discretized the equations with respect to space, but the time variable is still continuous.

To evolve the conserved quantity, we need to integrate the semidiscrete form (11) with respect to time. We can use any standard method for solving ordinary differential equations (ODEs). As we have used a first-order method in space, we choose to use a first-order method in time. Note that a higher order method in time is not needed because we have used a first-order method in space. In this paper, we implement the first-order Runge-Kutta method (the Euler method) to integrate the semidiscrete form (11) with respect to time. The fully discrete finite-volume method is

$$\mathbf{U}_j^{n+1} = \mathbf{U}_j^n - \frac{\Delta t}{\Delta x}(\mathbf{F}^n_{j+\frac{1}{2}} - \mathbf{F}^n_{j-\frac{1}{2}}). \tag{12}$$

The flux is approximated using the Lax–Friedrichs formulation (see LeVeque [5] for more details):

$$\mathbf{F}_{j+1/2} = \frac{1}{2}(\mathbf{F}_j + \mathbf{F}_{j+1}) - \frac{\Delta x}{2\Delta t}(\mathbf{U}_{j+1} - \mathbf{U}_j). \tag{13}$$

Here we have dropped the superscript n for brevity of writing. Considering the Courant-Friedrichs-Lewy (CFL) condition, we do the numerical computation in such a way that the numerical method is stable.

4. Numerical results

In this section we present some numerical results to show that the numerical method works well for solving (1) and (2). All measured quantities are assumed to be in SI units with the MKS system.

Consider a circular pipe with $L = 1200$, $D = 2$, $c = 1000$, and $g = 9.81$. We assume to have the initial conditions

$$H(x,0) = \begin{cases} 10.00, & \text{if} \quad 0 < x \leq 1000, \\ 42.44, & \text{if} \quad 1000 < x \leq 1200, \end{cases} \tag{14}$$

and

$$Q(x,0) = \begin{cases} 1, & \text{if} \quad 0 < x \leq 1000, \\ 0, & \text{if} \quad 1000 < x \leq 1200. \end{cases} \tag{15}$$

The boundary conditions at $x = 0$ are

$$H(0,t) = 10, \quad Q(0,t) = 1, \tag{16}$$

and at $x = 1200$ are

$$H(1200,t) = 42.44, \quad Q(1200,t) = 0. \tag{17}$$

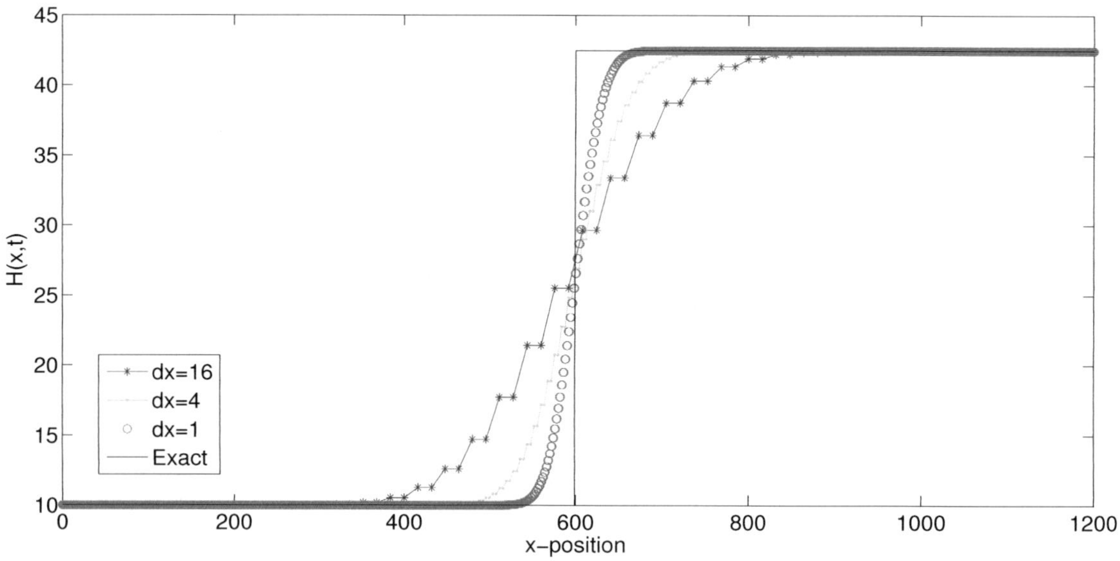

Figure 1. Results for varying cell width Δx at time $t = 0.4$ with dx meaning Δx. Here $\Delta t = \Delta x/2c$. The numerical solutions converge to the exact solution as Δx approaches zero.

With these settings, we simulate the problem to obtain numerical solutions at time $t = 0.4$ and analyze the results.

Based on simulation results, the presented numerical method is diffusive. However, the diffusion can be reduced by taking more computational cells, that is, taking a smaller cell width Δx, as shown in figure 1. (To generate this figure, we take $\Delta t = \Delta x/2c$ for our numerical simulations.) This could make the computation more expensive, but in fact it is not too expensive due to the nature of the first-order numerical method and the model. We observe that taking smaller Δx leads to more accurate numerical solution. Therefore, by numerical experiments our numerical method converges to the exact solution, as the cell width tends to zero. We observe that taking the time step $\Delta t > \Delta x/c$ results in an unstable method.

5. Conclusion

We have proposed the use of the conservative finite-volume method to solve the water flow problem in a pipe. The method is simple in structure, because it is explicit. The method is conditionally stable. Our results are limited to one-dimensional problems. Possible future direction may extend these results for multidimensional problems.

Acknowledgments

This research was financially supported by Sanata Dharma University and a research grant from *Direktorat Riset dan Pengabdian Masyarakat* (DRPM) of Ministry of Research, Technology and Higher Education of the Republic of Indonesia year 2016.

References

[1] Chaudhry M H 1987 *Applied Hydraulic Transients* (New York: Van Nostrand Reinhold)
[2] Mungkasi S 2016 *AIP Conf. Proc.* **1737** 040002
[3] Škifić J, Maćešić S and Črnjarić-Žic N 2010 *J. Hydraul. Eng.* **136** 483

[4] Mungkasi S 2016 *Adv. Math. Phys.* **2016** 7528625
[5] LeVeque R J 2002 *Finite Volume Methods for Hyperbolic Problems* (Cambridge: Cambridge University Press)

Institute of Physics
Dirac House, Temple Back
Bristol BS1 6BE UK

ISSN: 1742-6588
ISBN 978-1-5108-4192-5

CURRAN ASSOCIATES INC.
proceedings.com